**Self-healing Control Technology
for Distribution Networks**

Self-healing Control Technology for Distribution Networks

Xinxin Gu
NARI Technology Co. LTD

and

Ning Jiang
Nanjing Power Supply Company
China

with

Kan Ji, Huiyu Li, Xinhong Qiu, Weiliang Li, Xuejun Ji,
Hongwei Du, Bingbing Sheng, Hai Huang
NARI Technology Co. LTD

Registered Offices
John Wiley & Sons, Inc., 111 River Street, Hoboken, NJ 07030, USA
John Wiley & Sons Singapore Pte. Ltd, 1 Fusionopolis Walk, #07-01 Solaris South Tower, Singapore 138628

Editorial Office
1 Fusionopolis Walk, #07-01 Solaris South Tower, Singapore 138628

For details of our global editorial offices, customer services, and more information about Wiley products visit us at www.wiley.com.

Wiley also publishes its books in a variety of electronic formats and by print-on-demand. Some content that appears in standard print versions of this book may not be available in other formats.

Library of Congress Cataloging-in-Publication Data

Names: Gu, Xinxin, 1952– author. | Jiang, Ning, 1965– author.
Title: Self-healing control technology for distribution networks / Xinxin Gu,
    Ning Jiang.
Description: Singapore ; Hoboken, NJ : John Wiley & Sons, 2017. | Includes
    bibliographical references and index.
Identifiers: LCCN 2016038230 (print) | LCCN 2016054149 (ebook) | ISBN
    9781119109334 (cloth) | ISBN 9781119109358 (pdf) | ISBN 9781119109365
    (epub)
Subjects: LCSH: Wireless communication systems–Quality control. | Wireless
    communication systems–Automatic control. | Computer networks–Management.
    | Self-organizing systems.
Classification: LCC TK5103.2 .G79 2017 (print) | LCC TK5103.2 (ebook) | DDC
    004.6–dc23
LC record available at https://lccn.loc.gov/2016038230

Cover Design: Wiley
Cover image: Michael Siward/gettyimages

Set in 10/12pt Warnock by SPi Global, Pondicherry, India

Printed in Singapore by C.O.S. Printers Pte Ltd

10 9 8 7 6 5 4 3 2 1

Contents

Foreword

This book has its origin in an R&D project designated by the Science and Technology Department of the State Grid of China – that is, the self-healing control system for urban power grid – and draws on our experience in the process of research.

The idea of a self-healing project began in late 2006, when final conclusions and reflections were made after the first trial of distribution automation. Many attempts have been made at large-scale pilots of Chinese distribution automation, from the end of the last century to the start of this century. Owing to the undeveloped nature of technologies, the successful cases were few. So, domestic electric power enterprises have not been motivated to carry out distribution automation, and all chose to reflect on and watch the development trend. This was a low-tide period for the development of distribution automation, and R&D on automation technologies and products remained stagnant.

However, during the *11th Five-Year Plan*, as China's economy grew rapidly, there were increased demands on power load intensity, complex power supply modes, and reliability. The demand for electricity could not be satisfied, due to the development level of the distribution network at that time.

The Technology Department of the State Grid Corporation of China, with the support of head office, implemented urban network planning and renovation, in order to promote coordinated development between power grid and cities, aimed at exploring solutions to problems existing in the distribution network and automation system, and with a new direction for distribution automation – especially for urban power grids.

During the project research, China suffered a serious and rare natural disaster in 2008 in the form of the Sichuan earthquake, which covered large areas of China, lasted for a long time, and caused great losses. In areas worst hit by the disaster, towers of the backbone transmission lines fell over and the lines broke, leading to water and power outage across nearby cities. The problems that the power grid faced suggested defects existing in the urban network back then. Once there is a breakdown in large-scale power grid under extreme weather conditions, power outage will spread across an entire area. Urban power grids bear most of the power loads of cities, so national political and economic security are involved, as well as the lives of hundreds of thousands of households. Once large-scale power failures occur in an urban power grid, many areas will be affected and suffer from great losses. The grid structure, power source structure, and corresponding distribution automation are essential, and research on self-healing systems has become increasingly important.

Research on a self-healing control system project for urban networks was ended at the time the State Grid Corporation of China launched the smart grid in 2009. So, the members of the research group had the opportunity to participate in smart grid research, and this book was written with the aim of offering a platform for smart grid research and discussion both at home and abroad, exploring self-healing theory and methods in practice.

Our thanks go to the State Grid Corporation of China and the State Grid Electric Power Research Institute for their involvement and support.

Following expert advice, this English version of the book has added Chapter 5: Distribution Network Communication Technology and Networking (invited contribution by Mr. Hai Huang) and Chapter 9: Development Progress of Smart Grid in the World.

Xinxin Gu

Preface

China currently has the world's biggest energy network and energy utilization terminal, and the next two decades will witness a rapid development of industrialization and urban sprawl. Demand drives the development of the Chinese smart grid, as well as new developmental patterns such as energy peak-load regulation, new energy storage systems, and cyclical utilization of resources, smart management of energy utilization terminals, and intelligent network services.

Energy and the environment lie at the heart of world challenges in the 21st century. To respond to such challenges, it is a common choice for all countries to seek to tap and utilize renewable resources. Against the current backdrop, the power industry came to realize that new energy access is getting more and more difficult if it is based simply on traditional and conventional technologies and measures. The Chinese power industry faces new challenges, and the contradiction between electric power development/ resources and the ecological environment is increasingly serious. China will stay committed to reaching the twin goals of non-fossil energy accounting for 15% of primary energy consumption and emissions of carbon dioxide per unit reducing by 40–45% by 2020 compared with 2005, and further seek to develop green energies such as hydroelectric energy, wind power, and solar energy. Smart grid is the only solution to the large-scale development of resources, such as wind power and solar energy, and ensures the large-scale access, transmission, and absorption of renewable energies over long distances, distributing energy nationwide in a more efficient way.

The Ministry of Science and Technology, in its *12th Five-Year Plan* for the industrialization of major science and technology, has demonstrated clearly that its overall goal is to make great breakthroughs in core technologies, such as large-scale intermittent new energy grid-connection and energy storage, smart distribution and utilization power, large-scale grid smart dispatching and control, intelligent equipment, establishing a technology system and standard system for intellectual property protection, establishing a complete production chain, basically completing a smart grid featuring informatization, automation, and interaction, and upgrading the Chinese power grid to be efficient, economic, clean, and interactive. During the timescale of the *12th Five-Year Plan*, China established a UHV backbone structure covering the main energy bases and load centers. At the same time, China worked to promote intelligence in all aspects of a smart electric power system and build a comprehensive and smart service network covering most areas of China.

Self-healing technology is one of the core technologies of smart grid, and has been the subject of intensive interest in the field of world power systems. This book was written based on the authors' own experiences from a self-healing control system project for an urban distribution network, and experience and practice in the study and construction of smart grid. The book is being published to advance the research and establishment of China's smart grid and to provide a reference for researchers, developers, and technicians on the subject.

Ju Ping
Hohai University

1

Overview

1.1 Proposal of Smart Grid

Since the 1980s, great changes have taken place in computer, information, and communication technologies. With the continuous penetration of new technologies and materials, the electric power system – which is considered a traditional field of technology – is now facing great changes. The great change that is going to take place immediately is generated by the demand of users, national security, and environmental protection. On July 8 and 9, 2003, more than 200 experts from American Electric Power Industry Equipment manufacturing company, academia, industry organizations, national laboratories, federal and state government agencies met for the National Electric System Vision Meeting in Washington D.C.

The topics below were discussed in the meeting:

1) goals for the years 2010, 2020, and 2030 to achieve the vision;
2) challenges that may be faced in achieving the above goals and vision;
3) research, development, and demonstration needed for facing challenges;
4) time schedules for research, development, and demonstration.

At that time, the challenge for the American electric power supply was that the aged power grid and electrical facilities failed to meet the demand for economic development and held the economy back. Particular attention must be paid to this serious problem.

Power transmission and distribution play a significant role in electric power supply, and the market and America's economy (to a value of 10,400 trillion USD) are greatly dependent on a secure and reliable electric power supply. With the ever-increasing requirements of users for electric power supply, considering national security, environment protection, and energy policies, it is necessary to set out a strategy for power grid reformation: integrated approaches for market, policy, and technology; electric power system research, development, and demonstration; policy analysis and modeling; and coordination between federal, regional, and state departments. The aim is to form a competitive North American electricity market through plan "Grid 2030," providing sufficient, clean, efficient, reliable, and affordable electric power at any time and in any place, leading to the world's most secure electrical service.

Self-healing Control Technology for Distribution Networks, First Edition. Xinxin Gu and Ning Jiang.
© 2017 China Electric Power Press. Published 2017 by John Wiley & Sons Singapore Pte. Ltd.

All the above services are provided, based on:

1) Power backbone network – to achieve power exchange between the US East Coast and West Coast.
2) Regional Internet power grid – a strong supplementary to the power backbone network.
3) Regional power distribution network – to implement power distribution.
4) Final US power grid – including communication and control systems.

The conference proposed combining the technology of power grids with up-to-date communication, control, and electronic technologies in order to establish a more intelligent electric power system and thus achieve a real-time self-healing power grid by 2030 [1, 3].

1.2 Development Status of China's Power Distribution Network Automation

The pilot work of power distribution network automation that started during the upgrading of urban and rural power grids at the beginning of 2000 lasted for about 2 years. As power distribution network automation equipment and communication technologies are immature, most systems failed to achieve as expected, and the development of power distribution network automation was still not clear.

With the rapid development of China's economy, however, problems such as high load density of power supply, complex power supply modes, and increasing requirements for reliability of power supply occurred in major cities. Urban power distribution networks have the most concentrated loads, which is of concern for the security of state politics, the economy, and every household; therefore, reliable and safe power supplies are required, especially in major cities. Once an accidental power failure occurs in a major city, great losses will be felt.

In order to change the situation of power distribution networks lagging behind, the State Grid Corporation of China (SGCC) organized and launched 31 major power distribution network projects and an international consultation for Nanjing and Qingdao power distribution networks. These projects are intended to speed up the pace of power grid construction combined with urban development and optimize power grid structure, build a robotic power distribution network, and increase the capacity and reliability of power supply.

Since 2003, we have been studying the application of project "Grid 2030" and new technology in traditional power industries, and what we can learn from this to promote the development of China's electric power, especially Advanced Distribution Automation (ADA) as researched by the US Electric Power Research Institute (Epri) R&D team, led by Frank Goodman.

ADA is the future development goal of power distribution network automation. It mainly studies and aims to solve the following problems:

1) improving reliability and power quality;
2) reducing operating costs;
3) researching the integration of power distribution networks and distributed power supplies;

4) researching the coordination between power distribution systems and demand-side systems;
5) shortening the time of power outage and recovery;
6) providing more options for subscribers.

The fields of technology involved in the project include:

1) design of new intelligent electronic devices (IEDs);
2) research on low-cost and multi-functional static switchgears, sensors, monitoring systems, fault prediction, etc.

Figure 1.1 shows the future intelligent power distribution network described by ADA. The automation system is composed of a synchronous satellite, communication, sensors, intelligent electronic devices, a control substation (local agent), and a distribution control center. The power distribution system is composed of a distribution substation, FACTS, IUT, and DER. It is an ideal, controllable, and adjustable distribution network with a complete system that provides industrial, residential, and commercial consumers with secure, reliable, and high-quality power supply [2].

1.3 Development of Self-healing Control Theory

- Can the faults of an equipment system, if any, be controlled or eliminated by itself during operation?
- Are the faults able to self-heal like human or animal diseases?

Under the guidance of system science, the "self-recuperating" treatment principle of modern medicine, including immunization, defense, compensation, self-healing, and self-adaptation, can be used as a guideline for research on equipment self-healing and its application.

Following the self-healing control principle of an equipment system that aims at prevention and elimination of faults, the electric power system is constantly summarized, perfected, and improved in practice to create a self-healing control theory of electric power systems.

Modern equipment is becoming increasingly large, high-speed, automatic, and intelligent. In particular, high-speed turbo-machinery, industrial pumps, fans, compressors, centrifuges, and other major equipments that are widely used in the petrochemical industry, metallurgy industry, electric power industry, non-ferrous metallurgy industry, and other process industries are closely linked to the process of production, and thus form a great system. Should a failure occur in such a system, major accidents and great economic losses may be caused. Since the 1960s, the international engineering science and technology circle has developed equipment monitoring and diagnostic technology; predictive maintenance and intelligent maintenance are gradually being introduced to industrial enterprises and an emergency stop interlock system is widely used, which has played an important role in ensuring safety in production and achieving practical results.

The purpose of self-healing control research is to change the traditional methods that depend merely on emergency stop and manual maintenance and troubleshooting. Different from device diagnostic technology and predictive maintenance technology,

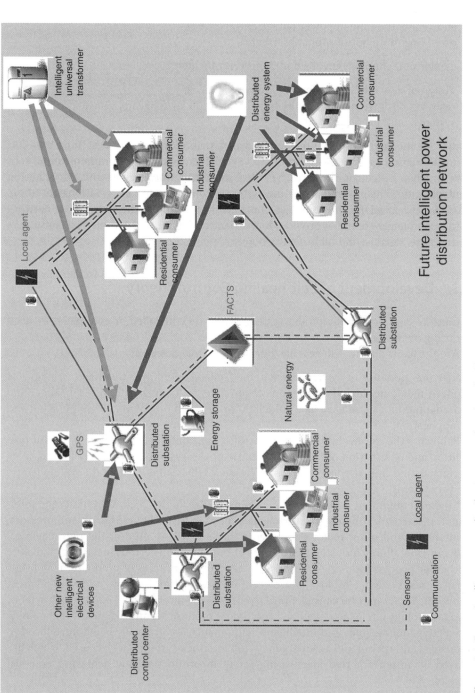

Figure 1.1 Future intelligent power distribution network as described by ADA.

this theory and method focuses mainly on how to provide the equipment system with a fault self-healing function and the ability to eliminate faults by itself during operation rather than employing manual troubleshooting with a mere emergency stop. Project practice has shown that the occurrence of most failures is a gradual progress, except for a few sudden failures. This means that equipment faults can be prevented by prompt and appropriate actions, once they are found at an early stage.

The occurrence of equipment failure is a gradual process. If the equipment is not designed with monitoring at important links, or it lacks intelligence, we may miss the opportunity to monitor and detect hidden dangers and these hidden troubles in the equipment will develop into failures.

Similar to an equipment system, an electric power system would first trip through the action of relay protection, followed by manual handling in the event of failure. An electric power system is a complex multi-device system and tripping can often cause multi-point disturbance; the possible consequences include large areas of disturbance or instability, such as splitting and generator tripping.

The concept of electric power system self-healing was first proposed in the Complex Interactive System Joint Research Project launched in 1999 by EPRI and the US Department of Energy. Later, the research projects Intelligrid of EPRI and Modern Grid Initiative of US Energy Laboratory both took self-healing as the main research objective, and deemed it the core technology for ensuring quality of power supply. The self-healing function has been a hot topic recently.

Self-healing refers to one function of a power grid that takes advantage of an advanced monitoring system to perform continuous on-line self-assessment of power grid operating conditions and takes preventative control measures so as to achieve timely detection, rapid diagnosis, rapid adjustment or isolation, with little or no human intervention. Remove the hidden danger and adjust the operating mode so that the failure can be isolated promptly upon occurrence, and the reconfiguration can be accomplished quickly and automatically; in this way the normal power supply would be unaffected, or affected as little as possible. Like the immune system in our human bodies, the self-healing function makes it possible for the power grid to protect against various internal and external damages (faults) to ensure secure and stable operation of the power grid and high-quality electric power.

The power transmission network is designed for a looped network and multiple feed structure, so that one or more components out of service won't affect the normal power supply of the system. Hence, the self-healing function on the one hand can achieve on-line monitoring of electronic equipment and find/remove potential faults by the removal of fault components in time through relay protection and on the other hand can perform on-line assessment of safety and warning/control so as to avoid widespread blackouts caused by power grid instability.

The power distribution network is user-oriented. Generally, power is supplied in a radiation mode. Any distribution network fault or power quality disturbance will have an effect on the quality of the power supply. Therefore, the self-healing function of a distribution network has some characteristics different from that of a transmission network. Self-healing functions of intelligent distribution networks include: firstly, reducing the duration and frequency of power failures, especially to avoid the problem of short-time unexpected power failures present in current power grids and increase the quality of power supply; secondly, optimizing the quality of power energy, especially restraining sudden drops in voltage; finally, effectively improving the ability to prevent disaster and damage.

It is necessary to note that the self-healing function is not a totally new concept in terms of definition and technology; relay protection and automatic safety devices both belong to it. The self-healing function is developed on the basis of traditional relay protection and automatic safety devices, but is more advanced. Its ultimate goal is to provide an uninterruptable power supply without human intervention.

The research and development/application and dissemination of self-healing control technology play a significant role in the construction of a smart grid and the improvement of power supply quality.

2

Architecture of Self-healing Control System for Distribution Network

2.1 Characteristics

The self-healing control system marks great new progress in the fields of network protection and control, and is designed to strengthen the abilities of self-prevention, self-adjustment, and self-recovery. It has two distinctive characteristics:

1) Preventive measures are taken to detect, diagnose, and eliminate potential failures in time. On-line monitoring technology is one such important measure.
2) There is a fault ride-through (FRT) capability.

Supported by the rich and comprehensive data available by real-time measurement and monitoring, the self-healing control technology motivates the power grid to operate through control measures, such as simulation technology, computation of short-circuit currents, protection-setting coordination, type identification, and load forecasting. Preventive control technology, used in the case of exceptions, prevents the system from failure and ensures the reliability of the system; the global control technique, during a fault, strives to avoid any power outage or shorten the outage coverage and time needed to restore power; the emergency control technology, in an emergency, would enter protection-acceleration procedures, initiate standby power, and turn the device from cool standby to heat standby. An emergency control pre-plan is a series of schemes arranged in several defense lines. The first scheme is carried out when the system is subject to the smallest disturbance; plans are worked out to prevent the system from entering into a fault state.

The final line of defense in emergency control is to recover control for the distribution network by self-healing control and relaying protection for the distribution network. Large-scale blackout will be prevented by improving the power output, cutting off part of the load, or even separating the whole distribution network into several temporary islands of power supply. The following technologies may be used:

1) A configuration and algorithm for optimum operation under normal conditions. When abnormally operated, the failure can be judged from the abnormal conditions or defects detected by an on-line measuring system of the equipment and distribution network. Later, the system will form a precaution strategy and take preventive measures. It needs to modify fixed values to relay protection on-line and check/simulate the modified values before the distribution network can enter a healthy state.
2) An algorithm and technology of prevention and control, an algorithm and technology of pre-reconfiguration for distribution networks. In case of failure, emergency control

Self-healing Control Technology for Distribution Networks, First Edition. Xinxin Gu and Ning Jiang.
© 2017 China Electric Power Press. Published 2017 by John Wiley & Sons Singapore Pte. Ltd.

measures will be taken, including isolating the faulty areas, reducing the blackout area and time, and reconfiguring the network to accelerate failure recovery.

3) A strengthening of the coordination among primary, secondary, and automation systems – applying the self-healing control technology to enhance the reliability of the distribution network.

4) A self-healing control system created all-round through coordination among global digital control, equipment on-line monitoring, relay protection for the distribution network and switchgear, and embedding advanced applications for self-healing control into the dispatching system. The self-healing control helps to ensure continuous power supply by reducing faulty areas and shortening any blackout time.

2.2   Structure of Self-healing Control System

The self-healing system consists of a self-healing control function, data interface, SCADA platform, and system platform, as shown in Figure 2.1. The self-healing system is embedded in a dispatching automation system by installing an additional self-healing control data server, system server, and workstation, as shown in Figure 2.2.

According to the security protection requirements for electric power, the electric power system is divided into four zones in order of importance of security. The self-healing system is at zone I and II, and interconnected with other systems by a physically isolating equipment and network firewall.

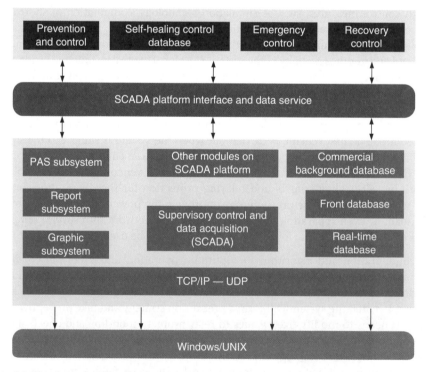

Figure 2.1 Structure of self-healing system.

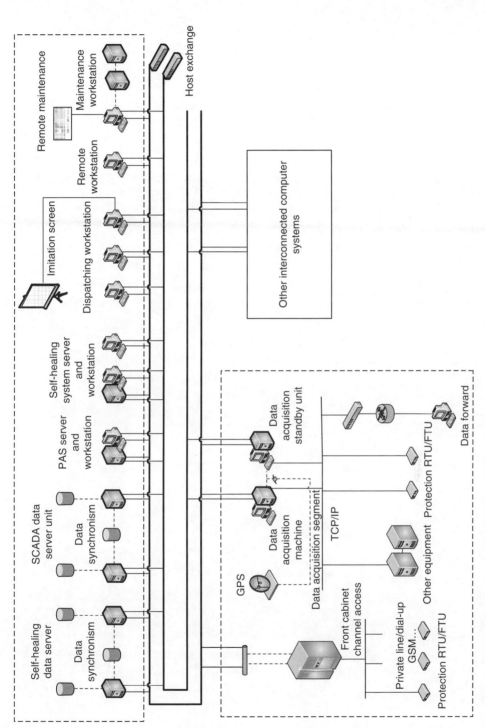

Figure 2.2 How the self-healing system is embedded in the dispatching automation system.

3

Advanced Application Software of Smart Dispatching and Self-healing Control for Power Distribution Network

3.1 Design Principles of Application Software for Smart Dispatching Platform

The smart dispatching platform for a power distribution network (namely the automation master station system of the power distribution network) is the main carrier satisfying various application needs in the operation, dispatching, and management of the power distribution network. It is structured with standard and general software and hardware platforms, and is properly provided with various functions based on the scale, practical demand, and application infrastructure of the power distribution network in each region. Meanwhile, the advanced system architecture is certainly proactive, so should be set up based on the principles of standardization, reliability, availability, safety, extensibility, and advancement, following the development direction of the smart power grid.

As one of the core technologies for smart dispatching of the power distribution network, the self-healing control software includes mainly the following application functions:

1) optimal operation control strategy in case of normal operation;
2) preventive control strategy in case of abnormal operation;
3) global control strategy in case of failure and coordination with relay protection;
4) emergency control strategy in case of emergencies (including the islanding strategy).

The application software of the automation master station system for the power distribution network is designed mainly on the following principles.

1) *Standardization*
   - Compliance with related international and national standards of software and hardware platforms, communication protocols, databases, and application program interfaces, etc.
   - Provision of an open system architecture and an open environment, suitable for stable operation on multiple hardware platforms and in such operating system environments as Unix and Linux, etc.
   - Compliance with IEC 61970 and IEC 61968.
2) *Reliability*
   - The software and hardware products incorporated in the system should be tested by the industrial certification authorities, to be assured of reliable quality.

Self-healing Control Technology for Distribution Networks, First Edition. Xinxin Gu and Ning Jiang.
© 2017 China Electric Power Press. Published 2017 by John Wiley & Sons Singapore Pte. Ltd.

- The key system equipment should be deployed in a redundancy configuration, and a single point of failure should not cause any loss of system functions and data.
- The system should be able to isolate the faulty nodes, fault removal should not affect the other nodes in normal operation, and the point of failure should be quickly recovered.

3) *Availability*

- The system software and hardware and the data information should be easy to maintain, with complete tools for testing and maintenance and diagnosis software.
- Each functional module should be flexibly configured, so that their addition and modification do not affect other modules in normal operation.
- The human–machine interface should be user friendly, with rich interaction means.

4) *Safety*

- Compliance with the requirements specified in the regulations on the Safety Protection of Electric Secondary System (SERC No. 5 Order) and the General Planning for Safety Protection of Electric Secondary System.
- There should be a complete rights management system to ensure information safety.
- There should also be a data backup and recovery system to ensure data safety.

5) *Extensibility*

- The system capacity should be extensible, to be added on-line with power distribution terminals, etc.
- The system nodes should be extensible, to be added on-line with servers and workstations, etc.
- The system functions should be extensible, to be added on-line with new software functional modules.

6) *Advancement*

- The system hardware should consist of mainstream products suitable for the industrial application trend, and meet the development demand in the power distribution network.
- The system support and application software should follow the industrial application trend and meet the development demand of application functions in the power distribution network.
- The proactive system architecture and design thinking should meet the development demand in terms of the smart power grid.

The supporting platform for the system software should comply with related industrial/ international standards; for the master station of a larger system, the server should use the Unix/Linux operating system; the network model should be the ISO-OSI 7-layered network reference model, and the network protocol should be TCP/IP-based.

The real-time database and the commercial database are combined together, which not only satisfies the real-time demand on the electric power system, but also reflects the superiority of the commercial database in the management and application of mass data. Standard API interfaces are provided for data access to multiple LANs/WANs, and for data interaction with other information systems (PMS, CIS, GIS, etc.).

Open functions on the application layer only involve those application programs that will not modify the original system configuration when the user's business process changes. With application extensions added to the parameter configuration, normal use of existing application programs and stability of system functions will not be affected.

3.2 Overall Structure of Automation System for Power Distribution Network

Besides providing the basic functionality for data interaction between the SCADA of the power distribution network and the superior dispatching automation system, the master station software of the automation system for the power distribution network should also satisfy actual demands in terms of the automation master station for the power distribution network, the application of related information management systems, and the automation research and development for the power distribution network, so as to build the application software functions stepwise in order when conditions permit. The overall structure of the automation system for the power distribution network is shown in Figure 3.1.

The automation master station system for the power distribution network is of a layered component structure, in which the distribution application can be completed on the heterogeneous platform through the bottom operation in the shielding layer of the application middleware. All the software modules must be designed in accordance with such international standards as IEC 61968, IEC 61970, IEC 61850, etc. to realize the standardization of data interaction and sharing with external systems and the plug and play of software products from third parties.

The supporting platform layer is the center of the entire system structure, the design rationality of which directly affects the structure, openness, and integration capability of the entire system. Through further analysis, it can be classified into three layers: integration bus layer, data bus layer, and public service layer. The integration bus layer provides a standardized interaction mechanism among various public service elements, application systems, and third-party software; the data bus layer provides the proper data access service for them; the public service layer provides various services for all application systems to complete, such as graphic interface and alarm service, etc.

3.2.1 Supporting Platform Layer

3.2.1.1 Integration Bus Layer

Compliant with open international standards such as IEC 61970 and IEC 61968, the integration bus layer provides a standardized interaction mechanism among public services, application systems, and third-party software, and is the integration basis between the system internals and third-party software.

The integration bus layer complies with the component principles specified in IEC 61970, and adopts advanced distributed object technology. The distributed object technology, represented by CORBA, innovates the software design method, as it allows objects to collaborate with each other and consequently constitute an organic integrity in the distributed object and heterogeneous network environments. Considering that the CORBA technology is characterized by supporting the heterogeneous systems, various programming languages, and integration legacy systems, it is the core of the integration bus layer for achieving the independence of software and hardware platforms of the integration system, independence of programming languages, position transparency, and ease of modification, maintenance, and migration, etc. The integration bus layer can support components of different granularity, which means that a component can be as large as a whole system, as medium-sized as an application

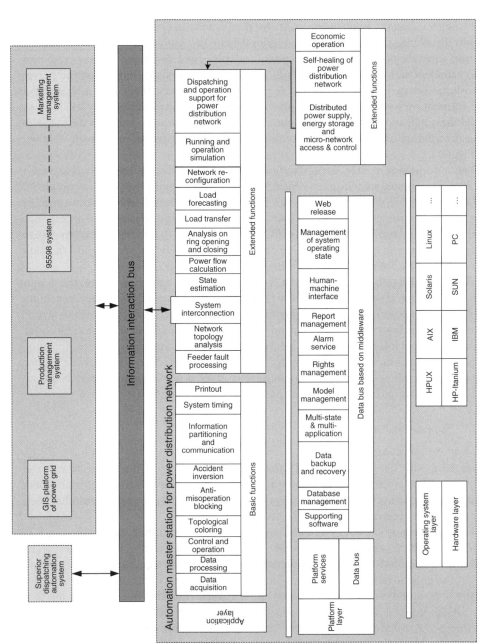

Figure 3.1 Overall structure of automation system for power distribution network.

system, or as small as a service element. Therefore, the integration bus layer not only supports integration with other independent systems from third parties and integration of third-party applications into the system, but also applies to integration among internal components in the system, so that all kinds of component are organically integrated into an entire system.

The integration bus layer also complies with IEC 61968 and establishes an information-based exchange mechanism. The integration targeted by IEC 61968 applies to independent application systems not internal system components, as it defines the interface reference model among systems and a complete set of message formats and modes. With the message broker and transfer functions implemented by the message middleware among different application systems, a loose coupling mechanism is provided for the independent application systems in a heterogeneous environment.

In conclusion, the integration bus layer plays a key bonding role which not only provides a standardized interaction mechanism between working service elements inside the system and various application systems and an effective mechanism for close integration of third-party software into the system, but also finds a reasonable way for the system itself to be integrated with other third-party independent systems.

3.2.1.2 Data Bus Layer

The data bus layer is composed of a real-time database, commercial database, and corresponding data access middleware. The commercial database is to store non-real-time and accidentally synchronized data and provide an historical data service. It is characterized by high reliability, big capacity, standard interface, and good safety. Relying on the bottom integration bus layer, it constitutes a distributed real-time database to ensure synchronization of real-time data.

3.2.1.3 Public Service Layer

The public service layer refers to various kinds of tool providing display, management, and other services to the application software. The public services focus on the general tools, whereas the application software focuses on the business solutions. Various kinds of application demand are sufficiently analyzed and summarized during the service design in order to set up a public service layer meeting all kinds of application demand.

1) Graphic tool: Provides graphic display/editing and graph–model–library integration functions.
2) Report tool: Provides a function for preparing various kinds of statistical report for applications.
3) Rights service: Provides the functions of rights management and assignment by the system administrator to all system users, and offers precise assurance methods for system rights management via a multi-level management mechanism consisting of functions, roles, users, and groups.
4) Alarm service: Processes various kinds of alarm/event and sends out the alarm information in a certain way as defined. Meanwhile, it separately records, saves, and prints out all the events, and provides retrieval and analysis services.
5) Web service: Provides an all-area web service, completely maintenance free.
6) System management: Includes system process management, redundancy configuration management, parameter management, resources management, operation monitoring,

etc., providing a complete set of management service mechanisms to help application systems implement their functions instead of self-implementing their own management mechanisms.

7) Analysis application service for power distribution network: Provides such analysis application component services for the power distribution network as network modeling, network analysis, power flow calculation, path searching, and load transfer.

8) Process service: Includes graphical process customization and drive engine of the business process flows.

9) Form customization: Provides a description of the individual business forms.

The process service and form customization mainly serve the dispatching and operation management of the power distribution network, so that the dispatching and operation management subsystem of the power distribution network can be realized based on the process drive and free form customization.

3.2.2 Application System Layer

The application system layer includes multiple external interfaces for SCADA, FA, GIS data conversion, and other advanced applications, all of which complete their own application functions with the support of integration buses, data buses, and public services, and are also organically integrated together as a whole system.

3.3 Smart Dispatching Platform Functions

3.3.1 Supporting Platform

The supporting platform provides a uniform, highly available, and highly fault-tolerant environment for the application software. The functions of supporting software include the integration bus layer, the data bus layer, and the public service layer. The supporting platform provides a standard development environment for users.

1) *Database management.* Not only provides easy-to-use and user-friendly maintenance tools of the database model and data as well as maintenance functions for users such as the addition, deletion, modification, and query of database table structures and stored data, but also supports database copying, backup, and recovery, as well as the management of distribution databases. As there are a lot of devices used in power distribution networks, they need to be quite carefully classified and optimized by data attributes, so that the database can be modified and queried conveniently and quickly. The database management function is available for all mainstream relational databases and real-time databases on the market, and is transparent for end-users.

2) *Graph–model–library integrated maintenance.* Supports graph–model–library integrated modeling and maintenance; supports the uniform modeling of real-time and future power grids; supports the partitioning maintenance of power distribution network models; provides distributed storage and uniform management functions for model information; provides verification and extraction functions for model information.

3) *Human–machine interface.* Provides the functions of picture editing, interface browsing, and interface management, as well as a supporting environment for application interface development and running. The picture editing function is to draw and manage primitives, graphs, forms, curves, and composite primitives; the interface management function is to customize picture style and menus; the interface browsing function is to browse real-time pictures, alarm information, and SOE information, and thus complete human–machine interaction.

4) *System management.* Monitors, dispatches, and optimizes system resources and can uniformly manage all kinds of application. It mainly includes the following functions: node management, application management, process management, network management, resource monitoring, clock management, backup, recovery management, and synchronization management of master/standby dispatching and operation controls, etc.

5) *Rights management.* Provides control methods for use and maintenance rights for various applications, which is an important tool for safe access management of application and data. This function provides the user with management and role management toward applications, so that multi-level and multi-granularity rights control can be implemented through the user's instantiation correspondence with the role. Support for the responsible area is provided to identify associated control between the user and the responsible area. The user-friendly rights management tool is also equipped to facilitate a user's rights setting and management.

6) *CASE management.* A public tool for the system to complete data storage and management on application scenarios, suitable for applications to conduct analysis and research with complete data under specific circumstances. Other functions include: CASE storage triggering, storage management, query, browsing, verification, and comparison, as well as CASE matching, consistency, and integrity check.

7) *Report management.* A mechanism providing report preparation, management, and query functions for all system applications to conveniently complete various report functions. It has report change and expansion functions, and also supports the definition, query, and statistics of yearly crossing data, yearly data, quarterly data, monthly data, daily data, and time interval data in the same report.

8) *Model management.* Responsible for setting up and maintaining power distribution network models in the system, namely the model basis for the application analysis software and operation monitoring of the power distribution network. The master station system should support flexible modeling methods, so that the automation master station for the power distribution network can import modeling from external systems or complete self-modeling with graphic models or graph–model–library tools. No matter how an entire grid model is set up, only one data source should be recommended, which means that if the model for the power distribution network is imported from GIS/PMS, then its automation master station no longer maintains the model for the power distribution network. A schematic diagram of the model management framework is shown in Figure 3.2.

3.3.2   Operation Monitoring of Power Distribution Network

The SCADA of the power distribution network is the basic function of the operation monitoring system for the power distribution network, which monitors and analyzes the operating conditions of the power distribution network in real time, ensures the

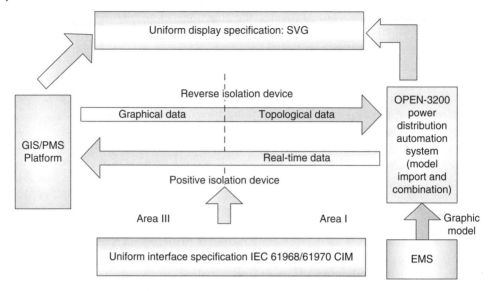

Figure 3.2 Schematic diagram of model management framework.

safety and economic operation of the power distribution network, and provides the basis for administrators at all levels to make decisions on production and management. The main functions include the following.

1) *Data acquisition and processing.* The system can acquire and process data of all types. All the acquired data has complete quality marks and source marks representing their states. All terminal data accessed by the system is periodically queried and acquired to ensure the real-time database. Data acquired from terminals and real-time data forwarded from the related automation system are immediately processed and stored in the real-time database. When they are processed and stored, their sources are transparent to users and application functions. Once processed, the data will become the basis and source of all application functions in the system. Besides data processing, all the data should also be labeled with quality tags to indicate reliability.

2) *Alarm processing.* This function provides abundant alarm actions and alarm behavior. Alarm actions include: voice alarm, audio alarm, picture alarm, printout alarm, Chinese SMS alarm, alarm to be confirmed manually, alarm on window, alarm library login, etc. Users can also self-define other new alarms.

3) *Partitioning and offloading.* The offloading technology for application information is used to effectively complete the offload and layered processing of all real-time information (full remote signaling, remote signaling COS, full remote metering, variable remote metering, plant operating condition, out-of-limit information, and various kinds of alarm information), reduce the message traffic flow on the network, and thus improve the response speed, the entire system performance, and the information throughput. Each center monitoring station only processes the information required to be processed in its responsible area, the alarm information about this responsible area is only displayed on the alarm information window, and the operations by dispatchers – such as remote control, number setting, blocking, and labeling – can

only affect the equipment in this responsible area, so that the information can be layered and isolated safely and effectively among different workstation nodes.

4) *System multi-states.* The system has multiple operating modes – real-time state, test state, and study state – for the debugging of terminal information and system functions without affecting the normal system functions. Besides, the system can be switched among the real-time operating state, the test state, and the study state.

5) *Network topology.* This is to complete the analysis and calculation related to the network topology, including topological applications such as electric topology analysis of the entire network, line coloring, power source point tracking, coloring of power supply areas, load transfer, etc.

6) *Anti-misoperation blocking.* Based on the conventional 5-protection/anti-misoperation principle, the topology-based 5-protection function should be provided as the system anti-misoperation blocking function based on network topology, giving good assurance on the safety of regulation and control.

7) *SCADA application based on geographic background.* The system supports the complete SCADA application based on geographic background, so that the dispatcher can conduct various kinds of SCADA operation on the geographic map and geographically locate the power distribution network equipment.

3.3.3 Information Interaction with Other Systems

Information on the power distribution network is extensive, and exists mostly in other related application systems. Therefore, the automation master station of the power distribution network needs to interact with other external information systems regarding the information to obtain more and more complete topology models, line graphs, and equipment parameters. According to IEC 61968, the interconnection and information exchange among multiple systems are generally implemented through an information exchange bus. For the data to be forwarded to/accessed from the dispatching automation (EMS), which is quite demanding in real time, direct interconnection may be adopted among systems. A schematic diagram of information interaction between the automation system of a power distribution network and external systems is shown in Figure 3.3.

In addition, when sharing information and integrating applications with other related systems, the master station system of the power distribution network strictly follows related national regulations on safety and protection, and the system safety also strictly complies with the requirements specified in the regulations on the Safety Protection of Electric Secondary System, the General Planning for Safety Protection of Electric Secondary System, the Safety Protection Planning for Secondary Systems of Dispatching Centers at Provincial Levels and Above, the Safety Protection Planning for Secondary Systems of Dispatching Centers at Prefecture and County Levels, the Safety Protection Planning for Substation Secondary System, the Safety Protection Planning for Power Plant Secondary System, and the Safety Protection Planning for Secondary System of Power Distribution Network.

1) *Interfacing with PMS/GIS.* This is completed via the information interaction bus. The PMS/GIS system exchanges graphic parameters of the power distribution network (single line diagram, liaison diagram, and geographic map) and other equipment information with the automation system of the power distribution network. The

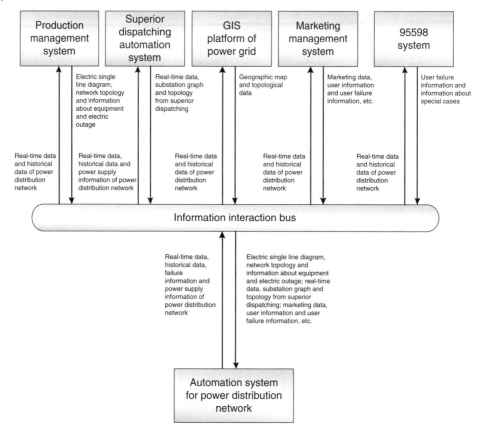

Figure 3.3 Schematic diagram of information interaction between automation system of power distribution network and external systems.

PMS/GIS system is the source of data entry to make the physical objects of the power distribution network on site consistent with the graphs of two systems.

2) *Interfacing with dispatching automation system.* The existing dispatching automation system completes the plant graph and model import interfacing with the dispatching automation system of the power distribution network via the computer network. Generally, single-way data transmission is based on the 104 communication protocol, in which the dispatching automation master station provides real-time information about the equipment on the substation side and related parameters for the automation system of the power distribution network.

3) *Interfacing with load control system.* Once the load control system receives the parameters of the power distribution transformers and remote metering data, the data about the power distribution transformers in the load control system is transmitted in E format in a reverse isolation way to the automation master station of the power distribution network, and the transmission frequency is determined by the load control system.

4) *Interfacing with "95598" customer service center.* The automation master station system of the power distribution network releases equipment outage information

and planned outage information on certain subjects and sends them to the call center via the information exchange bus. Then, the call center sends failure information on the subject cared for by the automation master station system of the power distribution network and transmits it to the corresponding adapter of the intranet master station system through the reverse isolation device, so that the dispatcher can analyze this failure information on the geographic map. This interfacing is required to be in accordance with IEC 61968.

5) *Reserved interfacing.* Considering the future development of the dispatching business and production management of the power distribution network, interfacing with other systems – such as a large-screen display, production commanding of the power distribution network, and power use marketing – is reserved to achieve single-way/two-way data transmission and the interaction of information flows and business flows among systems

3.3.4 Advanced Application Software of Self-healing Control

As the automation system of the power distribution network is characterized by massive scattering of multiple data acquisition points, with great variation in architecture for the power distribution network and frequent updates of equipment, it differs a lot from the automation system of the main network dispatching in terms of application demands, solutions, technical routes, and implementation processes. All the above differences are concentrated in the analysis application skills for the power distribution network, which thus directly affects the functionality implementation of self-healing control software.

The analysis application function for the power distribution network is the advanced application software for scientific management purposes, having particular importance for such smart applications as self-healing control, network optimization, and distributed power source access for the power distribution network. Therefore, the analysis application function should be gradually configured and completed in line with actual local demands and conditions during automation construction for the power distribution network.

The general configuration planning of analysis application software for a power distribution network is shown in Figure 3.4.

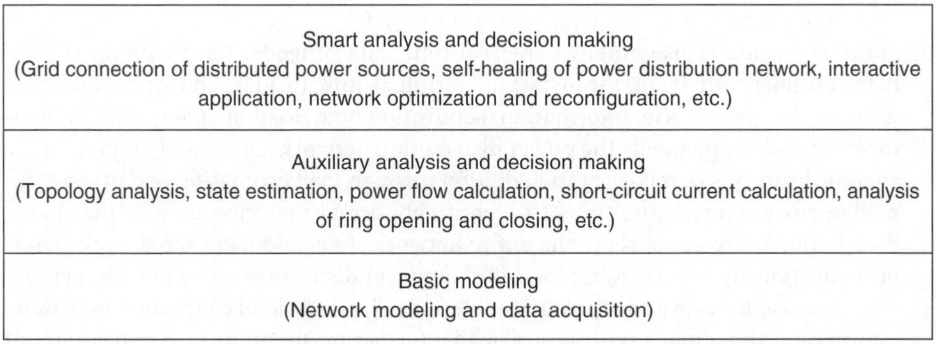

Smart analysis and decision making
(Grid connection of distributed power sources, self-healing of power distribution network, interactive application, network optimization and reconfiguration, etc.)
Auxiliary analysis and decision making
(Topology analysis, state estimation, power flow calculation, short-circuit current calculation, analysis of ring opening and closing, etc.)
Basic modeling
(Network modeling and data acquisition)

Figure 3.4 General configuration planning of analysis application software for power distribution network.

The basic modeling is to implement functions such as the automation of superior dispatching, geographic information acquisition, and graph model switching in the geographic information system and model combination and thus build up complete power distribution network models covering high and medium-voltage power distribution networks. Meanwhile, it is also to acquire the real-time operation data of the power distribution terminal grids on site, obtain the quasi-real-time data of the power distribution network from the marketing system via the information exchange buses, and provide the basic data profile for analysis applications for the power distribution network.

The auxiliary analysis and decision making includes such basic functions as the topology analysis, state estimation, power flow calculation, short-circuit current calculation, and analysis of ring opening and closing. With these analysis and decision-making functions, a consistency check analysis can be conducted for the operation profile of the power distribution network, so as to provide technical support for the basic auxiliary decision making in the dispatching of the power distribution network.

The smart analysis and decision making is mainly for the interaction and automation demands in the smart power grid. Considering the new characteristics and problems occurring in the operation of the power distribution network and the demands of smart decision making, the functions include the grid connection of distributed power sources, self-healing of the power distribution network, interactive application, and network optimization and reconfiguration, so that the future operation and dispatching of the power distribution network has smart features or elements.

The advanced application software of self-healing control is analyzed as follows.

1) *Topology analysis.* The topology analysis of the power distribution network is the foundation for all the analysis applications of the power distribution network; it is used to build up a dynamic model of the power distribution network. This kind of model shows the connection and link relations among devices and the real-time state of the power distribution network, which is suitable for any form of distribution network connection mode. Based on this model, the living area division and dynamic coloring can be conducted to analyze and identify the power supply points in the power distribution area and the power supply paths of all power supply points. The model structure varies with the devices shown in the wiring diagram of the power distribution network, and the model state also varies as the real-time information about the power distribution network is refreshed.

2) *State estimation.* This combines the real-time data of feeder terminal unit (FTU)/ data terminal unit (DTU)/transformer terminal unit (TTU), and quasi-real-time data in the power use information acquisition and load management systems together, and supplements the power distribution network data from comprehensive analysis by means of recovery and addition through load estimation and other compatible analysis methods, so as to identify abnormal metering data in the power distribution network, observe the entire network state, and meet the data demands of application analysis software for different power distribution networks. It includes static load calibration and topology calibration. The static load calibration uses static information and obtains static typical load information about the load nodes without measured values. Each load node can have multiple different load types calibrated for use. Once the static load calibration is done, the load node with the calibration

value of the static load obtained should undergo topology calibration – to correct the static load calibration value based on the topology structure of the network and the existing measured real-time value.

3) *Power flow calculation.* The node voltage, branch current, and power distribution of the entire power distribution network is calculated based on the topology structure under the specified operating conditions of the power distribution network, substation bus voltage (namely the feeder outlet voltage, support of city power distribution integration), and operating power of load equipment, etc.

4) *Analysis of ring opening and closing.* Ring opening and closing will cause a power flow change in the original power supply area. In order to maintain safety of power grid operation, it is necessary to calculate the power flow for ring opening and closing, so as to check whether the active power, reactive power, current value, and bus voltage of each related branch line are out of limits. As for the power flow for ring closing, the maximum impulse current and stable ring closing current in case of switching equipment closing are given, and the results of a safety check are analyzed. Through power flow calculation for ring closing, it is determined whether the ring closing current causes over-current protection or quick-break protection for the ring closing switching equipment misoperating or not, and also whether the ring closing is considered to cause over-current protection or quick-break protection for other system equipment misoperating or not. For the ring opening calculation, it needs to be determined whether the equipment with ring opened was overloaded.

5) *Feeder fault processing.* The automation master station of the power distribution network combines the fault alarm detected by each power distribution network terminal or fault indicator with the fault information about relay protection signals and circuit-breaker tripping of substation and switching station, and then starts the fault processing program and determines the fault type and position. The alarm can be in terms of sound/light, voice, or printout event. On the single line diagram of the power distribution network that is automatically pushed out, the fault section will be clearly shown through dynamic topological coloring. When necessary, the master station can provide one or more operation pre-plans for fault isolation and power supply recovery. The helping dispatcher conducts remote control to quickly isolate the fault and recover the power supply. The fault processing is considered on two: a simple fault and a complicated fault. If the ring network has dual power supplies and meets the $N-1$ principle, it means that when one power supply point malfunctions, the opposite power supply point can drive all the load on the ring network and the system will process the fault in the mode for processing simple faults. The schematic diagram of a hand-in-hand simple fault is shown in Figure 3.5. For the simple ring network structure, once the fault is confirmed, the capacity of the dual power supplies is first estimated and the $N-1$ principle is followed. In this case, the opposite power supply point can drive all the load on the ring network or the load in all non-malfunctioned areas downstream of the fault, and then the processing mode of the simple fault is available. For an instantaneous fault, if the substation outgoing circuit breaker is successfully reclosed and the power supply is recovered, the fault processing program will not start and only the alarm and related record will be given. For a permanent fault, once the substation outgoing circuit breaker fails to reclose, the fault processing program will start. If the ring network has multiple power supplies (more than two) or has dual power supplies but does not meet the $N-1$ principle,

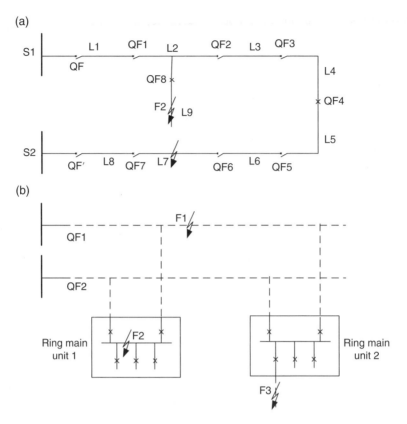

Figure 3.5 Schematic diagram of hand-in-hand simple fault: (a) overhead line; (b) cable line.

then the system will process the fault in the mode for processing complicated faults and the analysis software for complicated faults will need to be started.

6) *Smart operation ticket.* The existing operation ticket issuing modes can be classified for two kinds of operation ticket: smart and manual. Based on the topological connection relations among devices and the operating expert rules for system state equipment, the smart operation ticket is automatically issued through reasoning in which the details of primary, secondary, AC, and DC operation are all included. The advantage of the smart operation ticket is that it reduces the daily workload of the dispatcher. The operation ticket procedure depends on the electric topology connection relations, a great amount of experience, man-made rules, and usage habits. Different plants or stations will have different procedures as a result of different operating modes. The reasoning basis is that the plant wiring must be standardized, using recognizable standard wires. The system is developed with a vivid graphic user interface on which users can mouse-click the corresponding electric equipment on the graph to complete the ticket issuing process. Once every element is clicked, the corresponding operation type is selected to generate the corresponding operation steps to be simulated and implemented in the operation ticket. If the user clicks the equipment in an order that violates the electric work safety procedure or related limitations, the system will automatically give the alarm to prompt the operator

about the error instead of generating this error command. The rule base of the operation ticket describes a graph–ticket integrated basic framework that supports the entire process of operation ticket, including ticket issuing/generation and interpretation. Based on the limitation check and rights management mechanisms for network topological safety, the proposal, review, pre-issuing, implementation, and monitoring of operation tickets can be jointly controlled by human and machine, and the dangerous points can be accurately analyzed. For the use and maintenance of operation tickets, the operation ticket should be exported from the primitive in the main system wiring diagram for issuing, and no more secondary maintenance is needed.

7) *Network reconfiguration.* This actually refers to the optimization of the grid network structure. Generally, there are a certain number of sectionalized circuit breakers installed along the power distribution line and a few interconnecting circuit breakers on the trunk line or branch line end to obtain standby power. Under normal operating conditions, for purposes of economy and safety, information about the architecture of the power distribution network and the power use load should be comprehensively analyzed, and measures such as changing the operating mode of the power distribution network should also be taken to transfer the load among feeders for reasonable distribution.

8) *Load transfer.* When an accident happens or an outage overhaul area is set, the interval power outage may accidentally happen in the accident interval or the overhaul interval. Therefore, in this case, the load in these intervals needs to be transferred so as to minimize the outage range caused by accident or overhaul. For this reason, the load transfer function is needed. The load transfer function analyzes the affected load on the targeted equipment and then transfers it to a new power source point, thus providing a load transfer solution including the transfer path and transfer capacity.

9) *Simulation functions of power distribution network.* These can provide a realistic simulation environment for operators to complete the simulated operation and simulation training of dispatchers. The simulation operation of a power distribution network does not affect normal system monitoring, which mainly includes the following functions.

- *Simulation of control operation.* The control operation can be simulated on the substations, switching stations, ring main units, and circuit breakers. Besides, its operating interface is the same as for real-time monitoring and control. In the simulation state, the circuit breaker can be opened and closed freely on all computers for outage range analysis. Based on the tracking result of the outage area, the equipment failure causing the area outage is analyzed to directly reflect the condition of the simulated power grid.

- *Simulation of operating mode.* The operating plan under overload condition is simulated. With the smart self-healing function of the automation system for the power distribution network, and based on the line overload and predicted overload results, the operating mode is adjusted and calculated to simulate the manual/automatic handling method for accidents.

- *Fault simulation.* The simulation software can simulate all kinds of fault and system state change in any place, and thus provide solutions to actual operating models for reference. In the simulated study mode, any imaginary fault can be set artificially. The fault processing procedure automatically displayed by the system includes the fault positioning, isolation process, and preview of recovery strategy for the master station, etc. The simulation test environment equipped with feeder automation technology is available to simulate the faults, algorithms, and result matching sections in the primary power distribution network.

- *Static safety analysis.* Based on the structure and operating characteristics of the power distribution network, as well as the $N-1+1$ principle, the static safety analysis studies and consequently proposes quite complete safety indexes of the power distribution network that can quantitatively describe the severity of accident consequences, the entire network structure of the system, and the system safety strength. With these indexes, a firm foundation is laid for the static safety analysis of the power distribution network to implement the power supply capacity in the power distribution network area and the risk warning study.

4

A New Generation of Relay Protection for Distribution Networks

4.1 Principles and Application of Network Protection for Distribution Networks

The electrical power system is composed mainly of power generation, transmission, transformation, distribution, and utilization. Since the part of the power system with the most close and direct relationship with power consumers is closely bound up with people's lives and production, the reliable and safe operation of the distribution line will directly influence the majority of power consumers. A safe and reliable power supply is one of the main tasks for distribution network automation. Relay protection for a power system is mainly used to cut off the failure equipment in the system and ensure normal operation. With the protection function of the distribution line, it is possible to detect, judge, and isolate fault sections and restore the normal power supply for such sections, thus reducing the damage from faults. The relay protection for a power system should feature selectivity, rapidity, sensitivity, and reliability. Because the distribution network is an important part of the power system, its protection should also meet above-basic requirements [4].

Relay protection of a distribution network with newer and higher technical requirements is required for distributed power supply, micro-grid, and energy storage in distribution network demands. Meanwhile, with the development of the hardware technology for the protection device, protection algorithm, computer technology, and communication technology, various kinds of dynamic self-adaptive protection technology have emerged one after another, such as the self-adaption for protection constant value and protection mode, and so on. According to the development of the distribution network, power experts from various countries have put forward a succession of theories and algorithms to improve the relay protection level of distribution networks and optimize the management of distribution networks, thus realizing high-level automation of distribution networks.

Furthermore, the communication technology is one of the most rapidly developing technologies of recent years. The Ethernet is gradually replacing the field bus of industrial control. An Ethernet-based LAN has been established in many substations. Besides, in many areas, the SDH optical fiber ring has been laid among high-voltage substations, connecting the LAN of all substations together to form the WAN with related services. Then, communication standards and protocols are issued one by one (e.g., EPRI UCA2.0 and IEC 61850), so as to ensure information transfer in the substation and reliable

Self-healing Control Technology for Distribution Networks, First Edition. Xinxin Gu and Ning Jiang.
© 2017 China Electric Power Press. Published 2017 by John Wiley & Sons Singapore Pte. Ltd.

information exchange among substations. The appearance of a GPS-based phasor measurement unit (PMU) allows the wide-area protection to uniformly consider the data in all areas of the network, establishing a foundation for the wide-area protection of the power system. First, it can ensure the accuracy of the electrical quantities (e.g., voltage and current) collected in each node of the system, with error not more than 1 µs, which makes it possible to realize the accurate calculation of electrical quantities of the entire network at the same time. Then, the angles of the current and voltage phasor can be measured. Therefore, in theory, if a PMU is installed at each node of the whole network, all angle variables will become known quantities.

With the wide application of communication technology in power grids, the interaction and sharing of various kinds of power grid information, as well as the high rate and quick response of communication means, make the power grid protection move toward wide area and networking. At the same time, the protection mode and means become more and more perfect. It becomes a kind of development tendency in relay protection of the power system to use wide area and networking to realize self-adaption and self-learning of protection. Relay protection for the distribution network is also perfected and extended, along with progress in relay protection technology for the power system.

4.2 Adaptive Protection

As the first defendant line for power grid operation, relay protection is crucial to the safe, stable, and reliable operation of the power grid, the action of which shall be selective, rapid, sensitive, and reliable. Through the joint efforts of the R&D unit, design unit, manufacturing unit, and operation unit for many years, the correct action rate of the relay protection device has been significantly improved. In particular, the level of relay protection for ultrahigh-voltage and high-capacity equipment has been very high. However, due to limitations on the setting method and cooperation strategy of traditional protection systems, phenomena related to incorrect protection action often appear in previous power blackouts. The power failure in a large area is directly or indirectly bound up with the unreasonable constant value of the protection system. Then, the incorrect action of such protection further expands the blackout range. For a long time, power grid protection has adopted a set of fixed values to deal with various kinds of operation – i.e., setting the constant value of protection according to the maximum operation mode of the power grid to ensure the selectivity of protection and checking the sensitivity of the constant value as per the minimum operation mode of the power grid. However, since the setting and checking of a constant value are subjective and empirical to some extent, the actual method may exceed the defined range, making the constant value of protection unable to match the current operation mode. Hence, the best performance cannot be realized, thus delaying failure removal and enlarging the range of failure removal. Especially when there is a wide range of natural disasters, such as earthquakes, typhoons, heavy fog, and so on, many sections of the power network may suffer from tripping. In this case, the topology of the power network changes greatly, and the traditional constant value of protection cannot cover the new operation mode, resulting in refusal operation due to a serious decline in sensitivity or misoperation due to selectivity loss, thus further extending the range of large-area blackout.

In fact, with the increasing scale extension of power grids and the wide application of distributed energy sources, the structure and operation mode of the power grid become more and more complex, imposing a bigger challenge on the safe working of relay protection. Whether the relay protection can meet the necessary requirements has a direct relation with its collected information quantity, adopted protection principle, and operation criteria, as well as the correctness in setting the constant value. Traditional line protection refers to a set of fixed constant values, used to deal with various operation modes of power networks. So, when the operation mode alters, the constant value of protection is unable to adapt to such change. Furthermore, in most cases, the local information is used for setting of the constant value, causing many difficulties in setting calculation and setting verification. It is not unusual for international and domestic power outages to be accompanied by misoperation or refusal operation in protection. People have noticed the role of adaptive protection in blackout prevention, such as automatic adjustment of operation mode and protection constant value.

In addition, the development of communication and network technology makes it possible for power grid information to flow in the power network. With the high-speed and safe flow of power grid information, the relay protection device can obtain surrounding (even whole-network) information, which can effectively make use of as much power grid information as possible to realize the resetting of the constant value and self-adaptation to the operation mode, thus ensuring the best working conditions of relay protection [6, 7].

4.2.1 Development History and Features of Adaptive Protection

Since the traditional relay protection adopts a set of constant values to handle various operation modes, it is difficult to ensure the selectivity, rapidity, sensitivity, and reliability of protection action at the same time. Hence, the idea of adaptive protection was put forward in the 1980s, the essence of which is that the in-adaptive protection constant value can be timely corrected when the operation mode of the power network changes. In 1988, G.D. Rockefeller *et al.* [4] formally defined the concept of adaptive protection for a transmission line – i.e., it shall be able to realize the on-line adjustment of constant value, operation logic, and characteristics of operation according to the change in operation mode of the power grid. The article mainly analyzes the performance improvement of adaptive protection, pointing out that it is necessary to use a hierarchical system structure to realize the adaptive protection of the transmission line and gradually achieve the engineering application of an adaptive protection system through interconnection with a SCADA/EMS interface. In 1989, A.K. Jampala *et al.* [5] analyzed how to speed up the calculation of the self-adaptive constant value to meet the requirements of adaptive protection and designed a topology analysis program in a computer-aided protection system at the University of Washington, which predicates that it is really possible to realize the real-time setting of the adaptive constant value of protection by means of a high-performance parallel computing system and a rapid communication system through testing the calculation time of the constant value of self-adaption of one simulation system.

The traditional application system has a low communication rate because of the limited communication conditions. Before the 1990s, adaptive protection mainly referred to a protection device based on point or line information. Domestic and overseas experts at that time focused mainly on the adaptive setting algorithm based on local information. Some international scholars researched adaptive distance protection, which effectively extends the protection scope and accuracy of zones II and III, thus improving the safety and stability of the power grid. In the literature, adaptive distance zone-2 protection has been researched [6]. The result proves that adaptive distance protection can enlarge the protection scope of distance zone 2 and make it extend to a deeper range of the next adjacent line. A bus adaption based on graph-based connectivity theory was also proposed [7]. If using the adjacent matrix of a switch node to represent the complex in-station topology, the fault bus or circuit breaker can be accurately located and isolated. Domestic scholars also carried out some research on adaptive protection. Ge [8] gives an introduction to the effect and meaning of adaptive protection and forecasts its development prospects. Suonan *et al.* [9] put forward new operation criteria for adaptive distance protection, enlarging the operation range of distance protection and enhancing the operation sensitivity to zone fault on the premise that the distance protection malfunction is effectively prevented. Liu *et al.* [10] introduce the method of correcting the operation characteristics of adaptive ground distance protection, ensuring that protection can strongly resist the transition resistance and effectively prevent transient overreach.

With the rapid development of high-speed wide-area network technology and wide-area measuring technology in recent years, the low reliability and low speed of traditional communication has been overcome and the theoretical time for data updating of the dispatching terminal can be controlled to within milliseconds, making it possible to research adaptive protection based on whole-network information of the power grid. Consequently, this has become a public hotspot. For an adaptive protection system, if two power transmission lines are faulty at the same time within a period less than the calculation period of the adaptive constant value of the power grid, the adaptive constant value obtained from the last calculation will be ineffective for the fault a second time. Therefore, real-time setting of a protection adaptive constant value based on wide-area information, and providing a protection strategy when the real-time constant value cannot be realized, are key factors for adaptive protection. In order to improve the real-time capability of adaptive constant value calculation, it is necessary to speed up the topology analysis of the power network, calculation of a short circuit, and calculation of a constant value, thus making the whole setting calculation speed follow the change in speed of the power grid. This also becomes an issue of concern [8].

The wide-area measuring system (WAMS) technology based on time synchronization makes possible the acquisition and synchronization of whole-network information. Research on protection and stability technology based on WAMS whole-network information is also carried out at home and abroad. People begin to use multi-agent technology to solve the protection problem of a single device (e.g., a single line). However, this kind of application is still at a very low level. In most cases, only the simple concept of an agent is used. The wide-area adaptive protection of WAMS-based multi-agent technology deals with different information, and decomposes such information into different intelligent agents. In this way, the parallel processing of multiple agents reduces the calculation stress of the main core unit and shortens the decision-making time, thus enhancing the timely response ability of adaptive protection. Since the multi-agent technique is complex and immature, it is still necessary to carry out further research on the structure, communication mechanism, and control strategy of each protection agent.

4.2.2 Realization Mode of Adaptive Protection

Based on the adaptive protection function of a power system, adaptive protection refers mainly to the fact that the protection system can automatically adjust its operation parameters to respond to a status change in the power grid and keep its best function, as well as instantaneously altering the constant value, performance, and other related factors of the function according to the change in operation mode and fault condition of the power system. The best function here means the optimal state or characteristics of each protection unit in the protection system under the new condition of the power grid, which is required to ensure the best operation state of the equipment and the best stability of the power grid in the corresponding system. Meanwhile, it is necessary to consider the time response ability in the whole process.

According to its function and the range of information used, adaptive protection can be classified into self-adaption of the protection device and the protection system. Self-adaption of the protection device uses local information and can only adapt to a change in system operation mode over a small range. Then, self-adaption of the protection system uses system information through a communication network, adopts an adaptive protection algorithm to adjust the protection strategy, and is able to adapt to various kinds of operation mode change in the power system, thus ensuring the best protective result.

Self-adaption of the protection device refers to the fact that the relay protection device can change its setting value, operation characteristics, or protection logic instantaneously on-line according to the input signal or control action. Its adaption range is relatively small and usually limited to the local protection unit. In most cases, only the device unit itself adjusts the constant value of protection as per local information. So, this kind of adaption applies to the structural unit of the power grid.

Meanwhile, self-adaption of the protection system means that protection devices in the system can coordinate and cooperate with each other to ensure the optimal operation state of the power network and its equipment when the state of the power network changes.

Therefore, according to the function and range of self-adaption of the protection device and protection system, adaptive protection can also fall into local adaptive protection and area/wide-area adaptive protection. Here we also classify the adaptive protection from these two aspects. Local adaptive protection instantaneously adjusts the protection of the relay device to ensure optimal ability of the local area of the power grid, mainly according to the acquisition of local signals and other auxiliary signals. So, information collection is independent of communication channel (based on this form of information acquisition, the local adaptive protection system is also called non-channel adaptive protection). Then, area/wide-area adaptive protection refers to the fact that protection devices cooperate with each other to ensure optimal operation of the area or whole system. In this case, information is mainly collected and transmitted through a communication channel. The research object, thinking mode, emphasis, corresponding theoretical basis, and technical means of these two kinds of adaptive protection are also different to some extent. The former emphasizes the behavior of a single protection device, based on there being abundant information related to fault and grid operation state in the protection device signals. The theory is based on information engineering and the technical means are based on signal processing and analysis. The latter focuses on whole-system behavior, based on the relay protection devices in the protection system forming an organic combination through both mutual restriction and interrelation. The basic theory includes system theory, control theory, and information theory. The technical means refer to using systematic control strategy based on information and communication technology.

4.2.2.1   Local Adaptive Protection (Non-channel Adaptive Protection)

As early as the late 1980s, domestic scholars put forward the sequential quick action which refers to accelerating the protection action of the home terminal through the response of the transmission line to a circuit breaker. After the 1990s, domestic scholars successively proposed non-channel protection based on power frequency and transient fault information. Although they used different fault information (power frequency fault information or transient fault information), they also had common characteristics, such as only using single-terminal fault information, being able to protect the overall length of a protected element, and having quick-action performance.

At present, various kinds of longitudinal differential protection used in high-voltage lines have the advantage of quick action at full length. However, it must depend on communication and need the corresponding primary equipment or high-speed channel (e.g., optical fiber), which increases not only the cost but also the maintenance fee. In fact, even longitudinal differential protection needs cooperation from suitable backup protection to ensure the quick and reliable removal of a failure when the channel is invalid. However, there are lots of three-terminal lines in medium- and low-voltage transmission systems in our country. If a channel is not used in this case, then traditional three-section distance protection and current protection are mainly adopted. Such protection cannot complete a quick action when a terminal in the protection zone faults. So, for a power transmission line with power supplied at two terminals, if traditional three-section distance protection is used when a fault appears at the terminal of the local line, then distance protection away from the fault point can only cut off the faulty line through delaying section 2, which means there is a certain delay and the quickness requirement cannot be met well [9, 10].

With the development of communication technology, the communication channel of a power system has been established above the substation level and basically covered by an optical Ethernet. The sharing and exchange of information can be realized in a zone above the substation level, which provides abundant information on the power network for protection at all levels, allowing the relay protection (e.g., longitudinal differential protection) to act quickly at full speed and full length. Then, with cooperation from the backup and other protection, the correctness, reliability, and sensitivity of the relay protection is greatly enhanced. However, since the construction of a distribution network is complex, with enormous quantities and many aspects, huge investment is undoubtedly needed if the communication channel is laid for all distribution lines. Therefore, non-channel adaptive protection can still be used for power distribution.

Non-channel adaptive protection mainly refers to that using a line fault reflected by single-terminal electrical quantity to distinguish the fault when a communication channel is not required, thus realizing the full-length or sequent quick-action of the protected line and completion of fault isolation. It can be classified into power frequency non-channel detection technology and transient non-channel detection technology according to the kind of electrical signals used. Its biggest advantage is that the hypothetical communication channel is not required among FTU devices distributed along the feeder line, which saves investment costs, reduces the maintenance fee of the communication channel after system commissioning, and reduces the dependency of the whole distribution automation system on the communication system (ensuring the working mode with master and without standby), thus greatly improving the reliability of the whole distribution automation system.

In recent years, domestic non-channel protection for a distribution network with great influence mainly refers to the non-channel protection scheme for a distribution line proposed by Dong Xinzhou and other related researchers from the National Key Laboratory for Power Systems at Tsinghua University. This kind of non-channel protection judges the operation condition of a remote breaker by checking the current break caused by the operation of a remote circuit breaker and the time position of the current break, thus realizing relay protection for a power line which is independent of the communication channel, uses only a single-terminal electrical quantity, and features quick action and selectivity. The scheme has the following main features:

1) Rapid removal of a fault on the dual-power distribution line based selectively on the information that the current increases after the dual-power distribution line has faulted and the sound phase current disappears after operating the circuit breaker.
2) Since lots of distribution lines are branched, it breaks through the routine and uses the second break of the non-fault phase current and its time behavior to judge the fault section, thus realizing the rapid and selective removal of a branched distribution line.
3) Isolating a fault from both power supply terminal and load terminal based on the information that the current at the power supply terminal increases and the voltage at the load terminal decreases after the radial distribution line has faulted. Rapid removal of the fault on a radial distribution line, based selectively on the fault overcurrent at the power supply terminal and a break in the sound phase current after operating the terminal breaker, as well as the fault low voltage at the load terminal and a voltage jump after operating the circuit breaker.
4) For a distribution line installed with a single circuit breaker, the automatic detection of whether protection is at the power supply terminal or the load terminal according to the different directions of power flow for normal operation and fault operation, thus inputting the protection module in adaptive form.
5) As for the fault in a double-circuit line, it forms a protection logic according to exports of positive-sequence and negative-sequence elements, sending out enabling/blocking signals to an adjacent line as per protection logic and cooperating with distance protection zone 2 to jointly determine whether the circuit breaker shall be tripped, thus accelerating the fault isolation in distance protection zone 2.
6) For a fault transmission line, it proposes reclosing-based non-channel protection for the transmission line, realizing the instantaneous removal of a fault within the distance protect zone 2 of the transmission line. As for a fault outside zone 2, the power supply is recovered via reclosing; for a fault within zone 2 reclosing is blocked, thus isolating the fault rapidly.

4.2.2.2 Area/Wide-Area Adaptive Protection

For a long time, power grid protection has adopted a set of fixed values to deal with various kinds of operation – i.e., setting the constant value of protection according to the maximum operation mode of the power grid to ensure selectivity of protection and checking the sensitivity of the constant value as per the minimum operation mode of the power grid. Besides, the present setting principle for relay protection takes the electric element as a protection object and uses a local quantity as the basis of the protection operation, which only reflects the operation state of some point or small area in the power grid. So, the flow transferring after the equipment trips due to a fault can easily cause overload inter-tripping, although the overload tripping usually develops slowly and can be removed through alteration of the constant value of backup protection, as well as coordination of an automatic device (e.g., generator tripping, load shedding). Most conventional backup protection is composed of over-current protection or distance protection. The backup protection based on such an open protection principle has a relatively high rate of protection malfunction in the current operation condition of the power system. The paper "Research on WAN-based adaptive protection for power system"[1] includes the following description: according to the study on conventional backup protection and override tripping mechanism, it is considered that the non-unit principle used by conventional backup protection (e.g., the protection zone is open and covers multiple elements) is the main reason for override tripping of power systems. Furthermore, the conventional backup protection protects some local areas only based on locally collected data, which may spread to the whole network in serious cases, thus causing blackout of the whole network.

The main protection usually refers to the differential protection and high-frequency longitudinal protection composed of double-terminal or multi-terminal electrical quantities of the element line, which realizes full-line quick action within its protection zone. Then, the backup protection is achieved by multi-section distance protection and current protection based on single-terminal electrical quantities through stage difference coordination. With the constant development of the power grid, its network composition and current flow become more and more complex, which brings great difficulty to the constant value of backup protection and coordination of the time limit, finally resulting in a continuous delay of the time limit of the protection operation, which is bad for the rapid removal of faults (because rapid protection cannot be used as a backup protection outside its protection zone, and the backup protection is based on a cooperative scheme, especially for the distribution network). For example, in a ring main unit, the main load character point is usually used as the starting configuration point; the operation time limit of line distance protection is seconds (s); the time limit of distance protection of an adjacent line at a previous stage is added with one stage differential; its operation delay is calculated in seconds. In this way, the operation time of backup protection for the power supply is relatively long, which is not good for system stability.

Because of various problems and imperfections of traditional protection, large numbers of protection researchers at home and abroad began to study wide-area protection based on WAMS. It is a key factor for wide-area protection to research the

---

1 Peng, H., Research on WAN-based adaptive protection for power system, Southwest Jiaotong University, Chengdu, China, 2009, pp. 79–81.

algorithm, strategy, and system of relay protection based on wide-area information. With the protection of wide-area information, the real-time operation state of the system can be known. Various kinds of protection and protection stability control device cooperate with each other to solve the long-time relatively isolated situation, proving effective and safe stability control measures for interconnected power grids. Besides, the rapid development of computer and communication technology provides necessary basic conditions for wide-area protection, making the exchange of wide-area information possible.

Therefore, the adjustment of power grid operation mode, adjustment of protection scope, and change of power grid topology structure all need the replay protection to accordingly have dynamic adjustment and adapt to changes in the power grid. The acquisition and exchange of power grid information has become a precondition for adaptive protection.

The area/wide-area adaptive protection of a power grid uses information from just the area or the whole network to analyze the current operation state of the system and instantaneously set and adjust the constant value of protection to make the protection adapt to changes in the power grid and rapidly, reliably, and accurately remove faults. Since traditional power grid protection uses a set of fixed constant value and local quantities in most cases as operation criteria to deal with various operation modes of the power grid, area/wide-area adaptive protection can effectively cover the shortage of traditional protection mode and reduce refusal operation or maloperation of protection, thus avoiding blackouts in large areas.

Experts at home and abroad have mainly studied two aspects of area/wide-area adaptive protection: stability control and relay protection.

1) *Stability control.* The concept of wide-area protection was first proposed by B. Ingelsson *et al.* [11]. The wide-area protection in their paper is used mainly to prevent long-term voltage collapse; the protection refers to generalized system protection, rather than relay protection of the electrical element. Such a wide-area protection system provides voltage stability control, other than relay protection. It is established based on a SCADA system and adopts a centralized decision-making structure. Although the SCADA system collects non-real-time data and has a low refresh rate, the voltage stability control doesn't require rapid real-time data exchange. So, such data can meet the application requirements. Also, some scholars at home and abroad have placed the research emphasis of the wide-area protection system on the system protection and control means of conventional protection and SCADA/EMS. The wide-area protection is mainly responsible for the stability control of the system, including automatic reactive power control, automatic voltage-regulation control of the transformer, generator tripping, load shedding at low frequency and low voltage, remote load shedding, system splitting and FACTS, etc. Compared with traditional stability control strategy, the wide-area protection system involves broader geographical scope during signal acquisition, control strategy formation, and control measure execution, which needs more complex calculation. Besides, with the development of communication technology, the WAMS based on phasor measurement also attains rapid development. It can be used to achieve real-time and continuous measurement, monitoring, and event recording, thus realizing dynamic disturbance monitoring of the system. However, some problems still have

to be solved for WAMS, such as communication protocol, sensor, GPS time setting, construction cost, and so on.

2) *Relay protection.* Japanese scholar Yoshizumi Serizawa and others put forward the idea of using a signal to accurately synchronize time and transmit multi-point current information through a special optical fiber channel to form wide-area current differential backup protection, so as to solve the problem that the existing current differential protection of a single electrical element cannot provide rapid backup protection. Some scholars have suggested a wide-area current differential protection system based on a multi-agent, which realizes the dynamic on-line partitioning of current differential and backup protection areas with the help of an expert system, then achieves primary and backup current differential protection of the whole power grid through mutual cooperation of protection agents. As an important part of wide-area protection, wide-area relay protection plays a vital role in assisting system primary protection, improving the adaptive ability of the protection constant value, simplifying protection cooperation, and shortening protection operation time, which is helpful to fundamentally solve existing difficulties in relay protection (such as weak adaptive ability, complex setting cooperation, and so on) – thus improving the adaptive ability of protection. However, when the wide-area adaptive protection is used in practice, if centralized protection is applied to the whole system, a large number of data acquisition points, mass data, transmission distance, speed, and other factors caused by the increase in system scale will make realization of the wide-area adaptive protection – as well as configuration, operation, and mainte-nance of protection – more difficult. In this way, the reliability of protection is hard to guarantee. Therefore, it is necessary to combine with an actual system, ensure the synchronization of time and data of the WAMS system, and define a wide-area replay protection zone when wide-area adaptive protection is applied. In this regard, using a localized multi-agent system structure can be considered to disperse the computation amount of the system, adjust the backup protection reasonably, and provide a timely display function of traditional relay protection in the area or at the time when wide-area adaptive protection cannot be used, so as to ensure the adapta-tion and correctness of wide-area adaptive protection. Besides, how to realize the real-time setting of a protection adaptive constant value based on wide-area infor-mation and provide the protection strategy when a real-time constant value cannot be realized are also key factors for the realization of adaptive protection. That is to say, how to speed up the topology analysis of the power network, calculation of a short circuit, and calculation of a constant value (thus making the whole setting calculation speed follow the change in speed of the power grid) is the basis for instantaneous improvement in adaptive constant value calculation and realization of an adaptive protection system.

4.3 Networking Protection for Distribution Network

In recent years, large cities in many countries have suffered from widespread blackouts in succession, causing enormous economic loss and resulting in severe impact on the normal social order. So, the reliability of the distribution network in large cities has attracted great attention. Compared with the areas which suffer from widespread blackouts in the world, the power grid in our country also has some problems, such as weak structure of the distribution network, protection and control measures not meeting the reliability requirements of the power grid, and so on. The main reason for such problems is that the protection device cannot adjust the protection mode and constant value in a timely and dynamical fashion to adapt to structure changes in the power grid, due to the complex nature of such changes and a lack of information. Such an adaptive protection mode needs help from a high speed and broad communication network.

The development of communication technology makes networking protection possible. With the rapid development of computer hardware, the power system also demands increasingly higher requirements for microcomputer protection. In addition to the basic protection function, microcomputer protection must also have a large-capacity memory for long-term storage of fault information and data, be able to process data rapidly, have a strong communication capacity, be able to network with other protection, control devices and dispatching to share data and information on the whole system and network resources (thus making the microcomputer protection device have the same function as a PC). Therefore, the networking of a microcomputer protection device is fully feasible under current technical conditions. Furthermore, in the past, in order to meet the needs of measuring, protection, and control, the secondary voltage and current of the transformer and line of the substation all had to be introduced into the master control room by means of a control cable. In this case, a large number of control cables are needed, leading to high investment and a very complex secondary circuit. In networking cases, a computer device integrated with the above protection, control, measuring, and data communication can be installed locally beside the protected device of the substation, converting related analog quantities into digital quantities and transmitting them to the master control room through a communication network, which saves a lot of control cables, reduces fault nodes, improves the reliability, and reduces operation costs.

4.3.1 Concept of Networking Protection for Distribution Network

4.3.1.1 Networking Protection

In a digital substation, according to protocol IEC 61850, two or more intelligent electronic devices at bay level exchange information through a digital communication network. Such a cooperative protection function is called substation networking protection. The cooperative protection of a communication network is based on the development of computer technology, network technology, communication technology, and microcomputer protection technology. Its significant features include decentralized data acquisition, decentralized installation, decentralized function, standardized information model, networking information sharing, local control and processing, and independence of station control layer. The instantaneity, reliability, and safety of network communication are crucial for the execution of networking protection.

For substation automation, the real-time requirement for carrying out a control function through a local area network is usually defined as 4 ms. The Electric Power Research Institute shows that the modern 100 M or 1000 M Ethernet switching technology can meet such 4 ms requirement, no matter whether the 10 Mbit/s Ethernet is connected through a sharing hub or through a switch hub. From the perspective of the communication protocol, mechanism, and technology, IEEE 802.3x full duplex reduces communication conflict;

IEEE 802.1w priority queue ensures the punctual arrival of important information; IEEE 802.1Q VLAN partition reduces collision probability; IEEE 802.1w rapidly generates a tree protocol to establish a network-redundant structure, providing the ability for rapid recovery, reducing network bandwidth occupancy, and improving equipment response performance. Such communication protocols ensure real-time network communication. The reliability of a network refers to the network's ability for continuous working without fault. The reliability of the network can also be ensured through a highly reliable network topology structure, redundant link design, as well as verification and retransmission mechanism of the application layer, in addition to reliable network equipment. The safety problem of a network is mainly caused by its openness, freedom, and lack of boundary. The protection network of a power system is independent of the open and boundless network environment, becoming a manageable and controllable safe internal network.

Therefore, the development of computer technology, communication technology, and network technology, and ensuring network safety, all make protection able to realize a corresponding function via networking, such as line protection, transformer protection, bus protection, and so on. The greatest benefit from network protection is data sharing and a realization of longitudinal protection, which can be achieved at first only by high-frequency protection and fiber-optic protection. Furthermore, since the substation protection system collects the current of all circuit breakers and the voltage of all buses of the station, the bus protection can easily be realized with no need for an additional bus protection device.

Data or information sharing is realized through digitized and networking data acquisition, which effectively reduces equipment redundancy, making it more economical and realizing the protection function, which cannot be achieved by previous conventional protection. Compared with traditional protection, networking protection can obtain more complete information. Then, the fault identification and judgment based on more complete information will be more reliable, thus improving the stability of operation.

4.3.1.2 Area/Wide-Area Adaptive Protection Based on Networking – Networking Protection for Distribution Network

From the above, we can see that area/wide-area adaptive protection just refers to realizing information sharing and exchange among protection devices (in the area or wide area), establishing a cooperative protection mechanism for the area/wide-area by means of a scientific protection algorithm and principle, as well as data processing technology, and automatically adjusting the protection mode and constant value of the device or system, so as to adapt to changes in the power grid. The distribution network automation is designed to realize rapid fault location, isolation, and recovery. However, because of the complex distribution network structure, diverse wiring method, and broad region scope, the improvement in automation level of the distribution network is limited to some extent. With the perfection of the communication network and improvement of the communication rate within the distribution network, the area/wide-area adaptive protection extends to the distribution network, from the distribution network dispatching to the distribution network substation, and then to feeder line, thus realizing area/wide-area adaptive protection within the distribution network. On the one hand, local regional adaptive protection can be realized in or between distribution network substations, between feeder lines, and between substations and feeder lines. On the other hand, network analysis and coordination within the system can be realized through dispatching control of the distribution network, thus achieving wide-area adaptive protection with centralized control from top down. Therefore, the networking protection of the distribution network combines a local mode with a centralized mode and uses various kinds of advanced protection technology/method to dynamically adjust the protection device as well as the protection mode and constant value of the system, so as to adapt to changes in structure and power flow of the distribution network, thus realizing area/wide-area adaption of the distribution network. Thus, the networking protection for the distribution network is the area/wide-area adaptive protection based on networking. The distribution network based on such adaptive protection is capable of rapidly locating fault points, locally handling faults, reducing system impact, quickly recovering power, etc.

4.3.1.3 Distribution Network Automation System – Fundamental Framework of Networking Protection

The automation system of the distribution network is used to monitor and control the operation of the distribution network, interconnect the distribution network with the SCADA, allow feeder automation and analytical application of the power grid and related applications, and is mainly composed of the master station, terminal, and substation (depending on the actual functional requirements) of the distribution network, as shown in Figure 4.1. The master station of the distribution network is the core of data processing and storage, man–computer interaction, and various application functions. The terminal of the distribution network is an automation device installed on the operation site of the primary equipment and can be classified as per the application object. The substation of the distribution network is the interlayer equipment between master station and terminal, usually used for communication collection and also able to monitor areas as required. The communication channel is a communication network that connects the master station, terminal, and substation of the distribution network to realize information transmission.

The master station system of the distribution network in an automation system architecture of the distribution network is mainly composed of computer hardware, operating system, supporting platform software, and distribution network application software. It can be used to realize basic functions of the distribution network SCADA, processing of feeder faults, analytical application of the distribution network, intelligent application, and other related extended functions, and is capable of exchanging information with other related systems via an information exchange bus. The processing of feeder faults refers to cooperation with terminals of the distribution network to identify, locate, and isolate faults and automatically recover the power supply in zones without fault. The analytical application of the distribution network includes model import/merging of the distribution network, topology analysis of the distribution network, loop-opening power flow, load transfer of the distribution network, state estimates, network reconstruction, calculation of short-circuit currents, voltage reactive control, load predicating, and so on. The intelligent application includes self-healing control of the distribution network (including quick emulation and early-warning analysis), economic optimized operation, interaction with other intelligent application systems, and other intelligent applications of the distribution system.

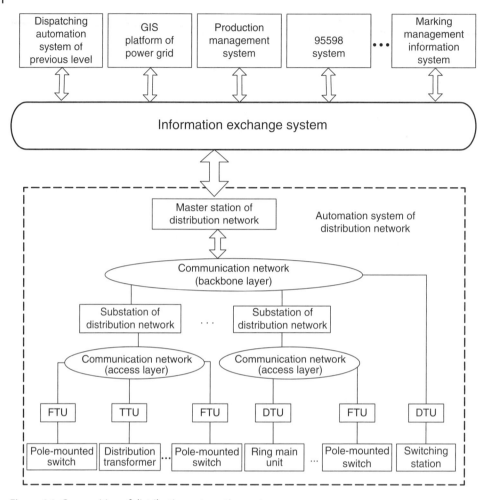

Figure 4.1 Composition of distribution automation system.

For the present distribution network, the application object of a distribution network terminal mainly includes switching station, distribution room, ring main unit, box-type substation, pole-mounted switchgear, distribution transformer, and distribution line. According to the object and function of application, the distribution network terminal can be classified into the FTU applied to the pole-mounted switchgear, the DTU mainly used in the distribution room, ring main unit (RMU), a box-type substation and switching station based on load switching, a TTU and fault indicator with communication function. Of course, some corresponding distribution terminal functions can also be achieved with other devices, such as a remote terminal unit (RTU), comprehensive automation device, or reclosing controller.

On the premise that the graph and topology source (dispatching master station, distribution network master station, power grid GIS, production management system, etc.) are unique, the master station of the distribution network can be used to realize interconnection of the distribution network SCADA and regional dispatching automation system. The distribution network SCADA can be used to realize data acquisition (support layered and classified collection), state monitoring, remote control, interactive operation, anti-misoperation locking, graphical display, event alarming, event sequence recording, post-disturbance review, data statistics, report printing, and operation status monitoring of the communication network of the distribution network, etc. Its interconnection with the regional dispatching automation system is mainly adopted to exchange data and share information with the regional dispatching automation system. Meanwhile, SCADA also follows principles for interconnection of the distribution network automation system with other systems.

The information exchange bus abides by IEC 61968/61970, which realizes information exchange between the master station and other systems through loose coupling. It supports the standard CIM model integrated with power generation, transmission, distribution, and utilization. The IEC 61968 information exchange model transforms a non-standard private protocol into a standard protocol via an adapter, thus realizing service-oriented architecture data integration.

With the help of an information exchange bus, it is possible to interconnect with external systems, integrate the distribution network information, extend the business process, establish a complete distribution network model, expand the application function of the distribution network automation system, and support locking management for distribution network dispatching, production, and operation, power consumption marketing, and other business. The access function of the distributed power supply/energy storage and micro-grid can be extended. The self-healing control and economical operation analysis of the distribution network can be realized through analytical application software of the distribution network, thus achieving upper-level cooperative dispatching with the power grid and interaction with an intelligent power system.

The measurement and control object of the distribution network automation system includes a medium-voltage substation, 10-kV switching station, feeder line, section circuit breaker, parallel compensation capacitor, user's power meter, and important load. The corresponding measurement and control functions are mainly executed by substations and terminals of the distribution network. The master station of the distribution network is capable of analyzing network topology, calculating loads, reconstructing networks, calculating power flow, dispatching and controlling, and so on. Meanwhile, it is possible to realize information exchange and sharing among different application systems through an information exchange bus interconnected with the upper dispatching system, a power utilization marketing management system, 95598 call center, and power grid GIS platform. With the development of communication technology, the communication network for a distribution network with several communication modes co-existing has emerged in the power distribution domain.

Therefore, the basic physical framework and functional framework for networking of the distribution network has already appeared among the master station, substation, and terminal of the distribution network, between the external application and the distribution network master station, as well as between the distribution network dispatching layer and the feeder execution layer. For example, the OPEN3200 master station system of the distribution network of SGEPRI has already been configured with an information exchange bus. The demonstration project of the research program Urban Distribution Network Automation System Based on Self-healing Control Technology has been carried out based on an OPEN3200 distribution network dispatching platform.

4.3.1.4 Networking: An Effective Method for Realizing Area/Wide-Area Adaptive Protection for Distribution Networks

Networking protection for distribution networks is designed to automate the distribution network, isolate faults, locate faults, recover the power supply, and maintain stable operation of the power grid. The whole process involves master stations, substations, and terminals of the distribution network. From a view of the distribution network automation, the most essential factor in power distribution is to realize feeder automation. Feeder automation adopts a local/centralized control mode and cooperates with outgoing lines of substations and switching stations based on the necessary primary framework, equipment, and communication of the distribution network to isolate faults and recover power supply, thus further improving the reliability of power supply. At present, feeder automation is mainly realized through the following two modes:

1) *Local control.* This does not need master station or substation control of the distribution network; realizes mutual communication, protection cooperation, or time sequence coordination via the terminal; isolates fault areas, recovers the power supply of a no-fault area, and reports the processing procedure and result when the distribution network suffers from a fault. Local feeder automation falls into the categories of recloser and intelligent distribution.
   - *Recloser.* A recloser is used to locally identify and isolate a line fault and recover the power supply for fault-free lines according to logic cooperation among the line breakers when a fault appears. This mode is suitable for a distribution line without communication means, or with a channel unable to be remotely controlled. However, when the primary framework of the distribution network is powered by a dual power supply, a local recloser is used.
   - *Intelligent distribution.* Isolate fault, recover the power supply for fault-free areas, and report the fault processing result to the master station of the distribution network through fault processing logic among distribution network terminals. The master station and substation of the distribution network may not participate in fault processing. However, a distribution line with peer-to-peer communication condition between distribution network terminals may adopt a local intelligent distribution mode.
2) *Centralized mode.* This judges the fault area, remotely or manually isolates the fault area, and recovers the power supply for fault-free areas by means of communication and cooperation between terminals and master station/substation of the distribution network when a fault appears. Centralized feeder automation can be classified into full-automation mode and semi-automation mode.
   - *Full-automation mode.* The master station or substation of a distribution network judges the operation state of the distribution network, centrally identifies and locates a fault, and automatically finishes fault isolation and power recovery in fault-free areas through the quick collection of information from the distribution network terminal in the area. For such centralized full-automation, the corresponding distribution line requires that there is a master–slave communication condition between the master station and substation of the distribution network; the terminal of the distribution network and the switchgear are configured with an electric operation mechanism.

- *Semi-automation mode.* The master station or substation of the distribution network judges the operation state of the distribution network, centrally identifies and locates a fault, and remotely or manually finishes fault isolation and power recovery in fault-free areas through quick collection of information from the distribution network terminal in the area. For a distribution line with channel unable to be remotely controlled, or switchgear not configured with an electric operation mechanism, the centralized semi-automation mode can be used.

In centralized and localized feeder automation modes, only the recloser mode can locally isolate faults based on a recloser and sectionalizer with no need for communication through reclosing the line repeatedly, avoiding outage of a full line due to a fault in some section of it. However, the local recloser mode also has the following drawbacks.

- The recloser can isolate a fault only through repeated reclosing, which has great impact on the distribution system and primary equipment.
- The power distribution for protection between reclosers in a loop circuit is realized through time delay. The more sections there are, the more difficult it is to realize cooperation of the protection differentials.
- In order to cooperate with the recloser protection differentials, the circuit breaker of a substation outgoing line realizes timed quick-break protection at the last level. The more sectional reclosers there are, the longer the definite-time protection delay of the outgoing line breaker is, which also has more and more of an effect on the distribution system.

Although non-channel adaptive protection technology has also undergone certain development, it needs support from various kinds of information and protection algorithm because of the variety of distribution network frameworks and wiring modes, and the complexity of faults. Besides, non-channel adaptive protection is also restricted by technology, especially the fact that the information extraction and protection principle based on transient components still has certain limitations. Furthermore, more and more distributed power supply is being added into distribution networks, which will alter the framework of distribution networks, making the electrical quantity characteristics of a system change greatly after a fault appears. In this case, the traditional fault detection method and relay protection mode will find it difficult to meet the safe operation requirements of a distribution network.

- The distributed power supply has various forms and the transient feature of a fault is totally different from that of a traditional generator, resulting in a complex breaking current feeding into the distribution network.
- Access to a distributed power supply makes the distribution network into a complex multi-power network. In this way, the power may be subject to bidirectional flow. The traditional current protection suitable for a single-power radial network will change the sensitivity, protection scope, selectivity, and other related aspects. So, the basic requirements for safe operation of a distribution network are difficult to ensure.
- The distributed power supply is greatly influenced by the weather and has various operation modes, making fault detection and protection setting calculations of the distribution network difficult.
- The current protection of a traditional distribution network doesn't usually involve directional protection. Access to a distributed power supply will change the direction

of power flow. In this case, the protection constant value, protection delay, and reclosing all have to be adjusted.

So, networking protection for area/wide-area adaptive protection based on regulation level, substation level, and feeder level of the distribution network is a good way to realize distribution network automation. It coordinates work between levels and dynamically follows, regulates, and optimizes the operation mode of the distribution network through regionalized and globalized information sharing. It also realizes feeder automation rapid fault isolation and power recovery of the distribution network, with help from the powerful calculation function of the computer at the dispatching level and based on network topology analysis, constant value calculation, load analysis, and coordinative protection at the substation level.

4.3.2 Realization of Networking Protection for Distribution Network

4.3.2.1 System Framework of Networking Protection for Distribution Network
Networking protection for a distribution network realizes information sharing or part sharing within the distribution network mainly through communication, and establishes information flow from the dispatching control to the switchgear based on an area/wide-area information network (although non-channel protection is a highly effective method, the protection principle based on fault transient quantity still has to be researched further, especially for distribution networks based on a complex framework and operation mode). From the process of networking realization, in addition to a communication network layer, it still includes a dispatching control layer of the distribution network, a substation layer of the distribution network, and a feeder line layer from top down. As a carrier for information flow, the communication network layer plays a role throughout the whole networking protection for the distribution network. The quality of communication channel, communicate rate, safety and stability of communication performance, and dynamic topology combination of the communication framework are directly related to the feasibility and safety of networking protection for a distribution network and the selectivity, rapidity, sensitivity, and reliability of relay protection; they influence the fault isolation and power recovery of distribution networks. The application of a dispatching layer and a substation layer of the distribution network in the networking protection introduced here is mainly realized by means of NARI OPEN3200 and DSA1000 programmable logic protection monitoring devices. The system framework of networking protection for a distribution network is shown in Figure 4.2.

4.3.2.2 Dispatching Control Layer of Distribution Network
As the central hub of a distribution network, the dispatching control layer of a distribution network takes charge of coordinating the operation and control of the distribution network, which is realized through the master station of the distribution network. The function of the master station has been introduced in previous paragraphs so we don't repeat it here.

4.3.2.3 Substation Layer
At the substation layer, the networking protection technology adopts a GOOSE mechanism in the digital substation to realize protection and control functions which cannot be achieved by a conventional comprehensive automatic substation

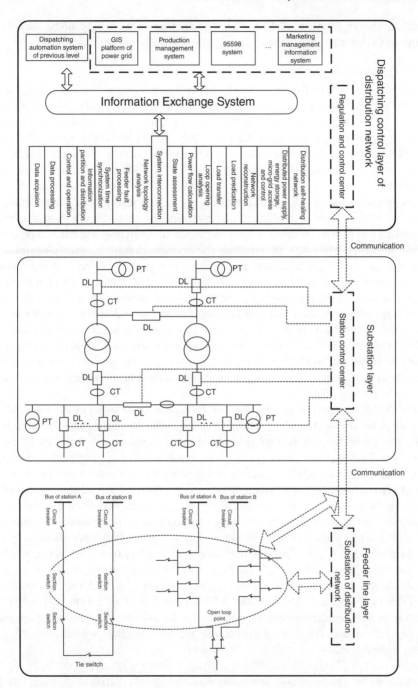

Figure 4.2 System framework of networking protection for a distribution network.

or be realized by special equipment, such as outlet protection, bus protection, backup power automatic switching, low-frequency load shedding, small current earth-fault line selection, anti-misoperation logic locking, and so on. We do not give any further details on the protection of the substation layer, as here we focus mainly on the realization of 10-kV bus protection, networking of backup power automatic switching, and current protection related to the substation.

4.3.2.4 Networking Bus Protection

A 10-kV bus fault is usually cut off by the backup protection at the low-voltage side of the main transformer. Since only the over-current backup protection, which needs delay before action, can process the short-circuit fault of a bus, this fault cannot be removed quickly, with bad consequences. In the past, we usually used bus differential protection to cover such shortages (i.e., introducing the current quantities of all loops at the bus section into a differential protection device – or differential relay – which however needs a number of secondary cables, thus resulting in increased construction costs. At present, using the GOOSE technology to establish a 10-kV networking bus quick-protection system can significantly simplify the secondary wiring. Compared with routine bus protection, networking bus protection can disperse the bus protection function into each bay protection unit on the premise that no hardware equipment is added and no exchange information is collected repeatedly. The GOOSE network instantaneously collects the fault information of each bay and sends related information to the equipment of the bay layer through the GOOSE service. Then, the equipment of the bay layer combines the identification result of a fault and operation mode to comprehensively judge a bus fault and send out a tripping command, so as to ensure the rapidity, selectivity, and reliability of bus protection operation. Since the GOOSE service supports the IEC 61850 communication protocol, it can realize networking of the process layer, which is an effective bus protection method in current digital substations, thus rapidly cutting off faults.

The 10-kV distributed bus protection is jointly completed by all protection and control devices for the outlet, sectional breaker, and inlet (including the protection and control devices at the low-voltage side of the main transformer) at the 10-kV related bus section. Each bay protection is only responsible for fault judgment and tripping outlet of the local bay, which is mainly finished by the configured power directional element. Then, the directional element sends out a GOOSE signal to lock the related protection. When a fault appears in the power grid, each dispersed protection unit (measuring and control device) will judge the direction of the fault current to see whether it is a bus fault, as shown in Figure 4.3. When the fault appears outside a bus (e.g., d3 malfunction), the fault current direction departs from the bus, the relay in the power direction of the fault near-end protection unit provides operation in the negative direction, and sends out a locking tripping signal to lock the bus protection. When the fault appears in a bus, the relay in the power direction doesn't provide operation in the negative direction, and does not send out a locking tripping signal. Each protection unit trips the corresponding circuit breaker without receiving the locking tripping signal.

The networking bus protection in a digital substation is based on the IEC 61850 communication structure and can easily be layered and distributed. Each bay protection unit can establish a peer-to-peer communication mode through the high-speed Ethernet, rapidly realize information sharing, instantaneously master the surrounding information of the substation, and find/distinguish an inside fault from an outside fault in a timely manner, so as to prevent incorrect tripping and ensure rapid response.

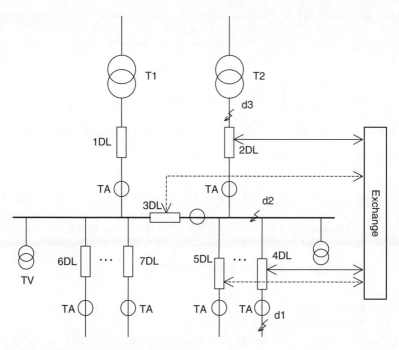

Figure 4.3 Networking bus protection scheme.

4.3.2.5 Network Backup Automatic Switching

Network backup automatic switching automatically switches the backup incoming line at the bottom layer of a GOOSE network using information such as the busbar voltage, incoming current, circuit breaker, and isolated switch position, collected centrally at the process layered network, combined with the current mode of operation, as well as the operation strategy, current state, and logical judgment, so that the measuring and control device can send orders to the process layer device, as shown in Figure 4.4. Compared with a conventional automatic switching device, it removes the line connecting the private auto switching device and protection and avoids repeated collection of information in a time interval. The transport process for network acquisition and transmission is simplified, which improves the reliability of auto switching. The action of automatic switching is applied to a transformer as well as the incoming line and bridge.

Figure 4.4 shows that the section circuit breaker 31QF is breaking, and incoming lines I and II are hot standby for each other (the case of incoming line I or II opening and section circuit break 31QF closing is not discussed in this section). In normal operation, 11QF/21QF closes and incoming line I/II supplies power to busbar I/II, respectively. When there is no load on busbar I, 11QF trips, 31QF closes, and incoming line II supplies power to busbar II. When there is no load on busbar II, 21QF trips, 31QF closes, and incoming line I supplies power to busbar I.

1) Charging conditions for standby automatic switching device (logic "and"):
   - 11QF closing position,
   - 21QF closing position,
   - 31QF opening position,

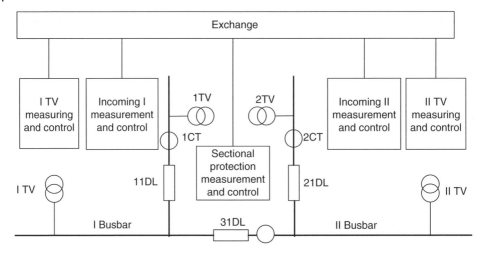

Figure 4.4 Networking standby automatic switching scheme.

- busbar I with voltage,
- busbar II with voltage.

2) Discharging conditions for standby automatic switching device (logic "or"):
  - 11QF opening position,
  - 21QF opening position,
  - 31QF closing position,
  - busbar I and II with no voltage synchronously.

3) Conditions for automatic switching bridge (logic "and"):
  - busbar II with no voltage,
  - no current in incoming line II (examine for no line current, otherwise include other external blocking signals),
  - busbar I with voltage,
  - 21QF closing position.

After switching the standby equipment automatically, 21QF trips, 31QF closes, and an action message is given out after a time delay. At the same time, it acts on the signal relay.

4) Conditions for automatic switching bridge (logic "and"):
  - busbar I with no voltage,
  - no current in incoming line I (initiate "circuit detector no current" and check the condition, otherwise check if there are other external blocking signals),
  - busbar II with voltage,
  - 11QF closing position.

After switching the standby equipment, 11QF trips after a time delay, 31QF closes, and at the same time acts on the relevant signal relay.

The logic relation of an automatic switching bridge is shown in Figure 4.5. A dispersed measuring and control unit will be responsible for discriminating BZT busbar I/II with or without voltage, incoming line I/II with or without current, and various TV break lines. The results will be sent to a main logic processing unit (or BZT device) in the form of a GOOSE signal via a communication network, which will process the data by the logic shown in Figure 4.5 and send it down to dispersed measuring and control

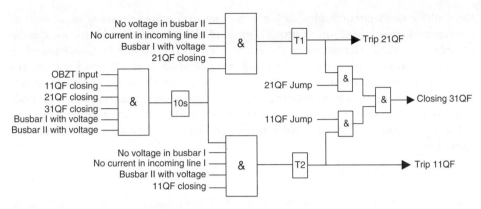

Figure 4.5 Logic Block Diagram for Automatic Switching Spare Bridge.

units in the form of a GOOSE signal via a digital transmission network. This is done to drive the smart switchgear to complete logical judgment. Among the data, analog quantity is transmitted to the measuring and control equipment using a sampled measured values (SMV) message in compliance with IEC 61850, while switching value and logic judgment are transmitted through a GOOSE channel. Owing to the network information mechanisms, whether the main processing unit (or BZT) is configured separately or centrally within the measuring and control device does not affect the operation of the BZT inside the substation.

4.3.2.6 Network Adaptive Current Protection

Current protection is the most common way to protect a distribution line. Currently, two-phase short circuits, grounding, and three-phase short circuits are usually protected by way of a three-sectional current. That is to say, current fast tripping (phase I), time-limit current fast tripping (section II), and over-current protection (section III) put together to form three-sectional feeder protection.

The traditional way to protect current fast tripping is by setting a three-phase short circuit at the end of the transmission line in maximum operation mode to ensure the selectivity of protection action. As for the sensitivity, it will be examined by the corresponding electric power in a minimum operation mode. This method of determining the protection set values ensures correct action of the relay in case of failure in all operation modes, but there are two shortcomings. First, the protection set values are not actually the best values in other operation modes; second, when the most serious faults occur in an electric power system in minimum operation mode, the relay protection system deteriorates, or even fails to operate. Owing to these shortcomings, the power grid is less flexible and reliable. A distribution system is composed of a large number of electrical devices, lines, and various users. Hence, the state of the system is in constant change (including load change and device switching). It is possible that various kinds of failure (transient, permanent, metallic, or non-metallic) will occur in the distribution network. Moreover, on some occasions, extreme operation modes may lead to system deterioration and failure to operate.

In any case, protection would not cut off the protected line by mistake. The set values should be set under any condition except for maximum short-circuit current in case of three-phase short-circuit faults. At the same time, to ensure sensitivity of protection, the range of circuit quick-break protection should be calculated under the most unfavorable conditions. That is to say, the scope of protection should be calculated when two-phase circuit faults occur in the minimum operation mode. Whenever there is a change in the mode of system operation or type of failure, the scope of protection will change. Therefore, the current setting value and time delay should be subject to mode of operation and state of failure, so as to improve the action of current protection.

The self-adaptive current for quick-break protection is calculated according to the formulation

$$I_{dz.1} = \frac{K_{re} * K_d * E_s}{Z_s + Zl} \tag{4.1}$$

where:

- K_{re} is a reliable coefficient of current quick-break protection, varying from 1.2 to 1.3. It is greater than 1 in consideration of the calculation error of the short-circuit current, error of relay current, non-periodic component, and necessary margins.
- K_d is the coefficient of fault types. It is equal to 1 in case of a three-phase short circuit and 3/2 in the case of a two-phase short circuit.
- E_s is the phase potential of the equivalent power source.
- Z_s is the actual impedance at the power side of the system.
- Zl is the impedance of the protected line, where l denotes the length of the protected line.

To set the values for current quick-break protection, we have to determine the type of failure (i.e., K_d) and the system-equivalent impedance (i.e., Z_s), which should be taken into consideration separately.

Generally, a distribution network operates in a radial network, or open ring operation mode. The long radial line on the basis of the outlet substation, together with outgoing lines from other substations, will form a dual-power looped network. The recloser or sectionalizer, load switch, and even circuit breaker on the feeders will form cascading protection. Communications permitting, the measurement signal, fault signals, and blocking signal of the directional unit on nearby devices are able to share information with each other and identify the mode of faults, breakpoint fault, operation mode of feeder, and grid structure of the distribution network. A double-end ring feeder will obtain fault signals on opposite sides of the circuit breaker through longitudinal differential protection to identify faults and realize quick protection. When communication conditions are not met, through the no-channel self-adaptive protection mentioned above, the system will use a positive-sequence, negative-sequence, and zero-sequence component for fault signal calculation to determine the mode of system operation, prevent maloperation of the circuit breaker in a non-fault zone, and accelerate tripping.

4.3.2.6.1 Feeder Line Layer The feeder line layer performs feeder automation, including fault detection, isolation, network reconstruction, and recovery, so it is vitally important to distribution network automation. It has many functions, such as monitoring the running state, remote control, or local autonomous control, fault isolation, load transfer, recovery, and voltage regulation. To isolate faults and restore power, the control system and switchgear coordinate with each other in order to identify, isolate, and restore faults in the shortest time, greatly reducing the time of blackout and improving stability of the system.

Despite the great differences in system performance due to different control modes, the general goal is to spend as little time as possible on execution, and restore as much load as possible. At the same time, load levels should be within limits and the voltage at the control node should be up to standard.

The following is a description of four methods to realize feeder automation. The factors that influence the choice of method are local condition, the status quo of the power grid and communication conditions, and one or a combination of methods used to protect the network distribution network. In the local mode, a recloser is used to realize feeder automation. Despite the fact that it is not in the scope of the distribution network in terms of concept and does not use a communication channel, a recloser will be described here as a method of feeder automation since it is locally adaptive.

1) *Local.* Includes reclosing and intelligent/distributed.

- *Reclosing methods.* 1. Recloser (circuit breaker) and sectionalizer. Reclosers are used in outlet circuit breakers and sectioners are installed at other cut-off points on the column. The recloser works in such a way that in case of failure, it will isolate faults by detecting voltage and time limit and closing the recloser at the upper level multiple times. Power is recovered by the order of the time limit. Communication means are not required, so it is simple. However, some problems still exist, for instance: multiple reclosing has an impact on the distribution device before faults are isolated; there is always a sectionalizer at one side connected to the faulty section, that needs to be open after the circuit breaker is closed; a fault may be identified by multiple reclosing in a non-faulty line. That is to say, non-faulty lines need to be out of service for a while; with a longer feeder and more sections, the time delay for each step will be longer, which has a great impact on the system. 2. Reclosers all used, including sectionalizer. A recloser is able to cut off the current of a short circuit and has the functions of protection and automation. When failures occur in a line, it will reclose the recloser at the faulty location multiple times and coordinate the time limits of protection action so as to isolate the fault(s) and restore power.

  Since communication means are not necessary in realizing feeder automation by reclosing, the recloser improves the performance, in contrast with a sectionalizer. This is because the use of a recloser prevents power outage along the lines caused by a fault in some sections and reduces the action time of outlet circuit breakers. However, there are shortcomings. Multiple reclosings are required before faults are isolated, which has a big impact on the primary device; when differential delay coordinated with protection occurs, more sections make it more difficult for the differentials to coordinate; the outlet circuit breaker is the last to have a time-limit quick break, so the more sectional reclosers and longer time delay of the outlet circuit breaker also have a large influence on the distribution network.

- *Intelligent distribution.* In this method, a load switch or ring main unit with motor-driven operating mechanism is used as section switch. At the same time, it is equipped with a feeder terminal unit that is capable of point-to-point communication. In case of line faults or outlet circuit-breaker tripping, the FTU device on the line switch equipment will exchange fault information via point-to-point communication (i.e., the FTU will pass on the message of faulty equipment to the FTU on a nearby switchgear, which will analyze, identify, and isolate faulty sections and restore power in the non-faulty area). Based on a communication network, the intelligent distributed feeder automation shares faulty tripping information with the distribution terminal at the nearby switchgear. It is also a kind of regional adaptive distribution protection based on characteristics of the distribution network.

There is no complete system to monitor local feeder automation by means of recloser and intelligent distribution, so it is local optimization of the distribution network. However, it can deliver outlet fault topology information to the dispatching layer via the outgoing circuit breaker, which is of great importance for network architecture, load analysis, load transfer, and even network protection.

2) *Centralized type.* Centralized feeder automation control is divided into full automation and semi-automation. The control model is composed of a distribution network, such as the outlet circuit breaker, FTU, DTU, communication network, or dispatching center. The FTU on each circuit breaker or main ring unit should communicate with the dispatching layer, and fault isolation should be under centralized control of the master station system for feeder automation in full automatic or semi-automatic mode (remote control/manual).

In case of permanent fault, the outlet circuit breaker of a faulty line substation will activate protection. If it fails to activate after the first reclosing, the central station computer will inquire (receive) information on the state of each FTU by polling. Fault detection software in the regulation and control center system will identify faulty sections based on the results of this analysis. The regulating and control center will give a series of commands to operate manually or with manual intervention, make proper arrangement of the network topology, isolate faults, reduce any influence on the network to a minimum, and finally restore the power supply to non-faulty loads.

Since the mode uses advanced computer and communication technologies, it can avoid the superposition of feeders and outlet circuit breaker and rapidly locate and isolate faults. At the same time, the time needed to isolate faults is not influenced by the distances between lines and the times of line sections. Since the centralized control is based on wide-area information, it is possible to restore power according to an optimal economic scheme. In addition, SCADA can be realized in normal conditions. Monitoring the condition of the feeder in real time and a "four-remote" function (YX, YC, YK, YT) meets the needs for normal operation.

3) *Localized and centralized integrated intelligence.* This is based on a localized and centralized smart feeder automation mode and a comprehensive embodiment of networked distribution network protection at the dispatching layer, substation layer, and feeder layer, as well as localized and wide-area adaptive protection. It is an improved control mode in the feeder automation system. In the control mode, remote control plays a major role, supplementary to local monitoring. It takes advantage of local monitoring and gives play to the strengths of the two control modes. As for the component of the control system, similar to a remote monitor, the outlet switching equipment uses a recloser and other line switches on the column use a load switch as sectionalizer. Each distribution network terminal is capable of communication. Moreover, the distribution terminal uses a self-adaptive smart algorithm (i.e., it realizes self-adaptive protection by use of a fault information

without channel protection technique). It works in such a way that it can realize localized control by giving play to the strengths of centralized control, with a complete channel, in case of failures in the communication channel, or if there is difficulty in remote control. That is to say, the intelligent algorithm will rapidly identify and isolate faults, and restore power. In so doing, it will solve the problem of lack of remote centralized control in case of communication failure and the control mechanism will be optimized, greatly improving the feeder automation system.

An integrated intelligent mode applies to places where there is a higher requirement for automation. When there is an optical fiber communication channel, use a circuit breaker or recloser on the line and a high-speed data channel for protection, and a supervisory mode in the optical fiber ring network. In this mode, there will be no time delay in the coordination of relay protection between switching devices. The integrated intelligent mode has the merits of point-to-point communication in a distributed localized control mode and remote control in a centralized control mode in the process of rapidly isolating fault sections, ensuring continuous power supply in non-faulty zones, and restoring power supply.

4.3.2.6.2 *Network Communication Layer* The network communication layer is an important layer in protecting the network, and the most basic structure to achieve network automation. It develops with the development of communication technology. The frame diagram of a communication network is shown in Figure 4.6. Here, the communication system is composed of an integrated access platform, backbone layer, and access layer.

1) *Integrated access platform.* An integrated access platform is configured on the main distribution network, which will perform integrated access of various kinds of communication mode, unified access specification, and management. The master station will be connected to the integrated access platform according to the interface specification. In addition, the communication access platform also serves other business and makes it unnecessary to build a separate communication system for a single business, facilitating the management and maintenance of the communication system. One favorable factor for the communication structure is that it is able to make the best of various kinds of communication means, and include more communication media for the sake of network protection. It also helps build up a bridge between the dispatching level and the feeder level to foster information communication.

2) *Backbone layer network.* A backbone communication network allows communication between master station and subsystem via an optical fiber transmission network; the information gathered by the subsystem will be accessed to SDH/MSTP or stored in an optical fiber network. In accordance with the information security standard, an IP virtual private network is used to realize the backbone communication network. The backbone communication network builds up a communication bridge via Ethernet for the dispatching level (master station) and transformer substation; since the former needs a large amount of information and data, high speed, high bandwidth, strong immunity to interference, and adaptation are at the center of information exchange. The topological matrix, results of network analysis, protection settings, and control commands are sent down to the substation level via a high-speed Ethernet, and perform adaptive protection for the substation.

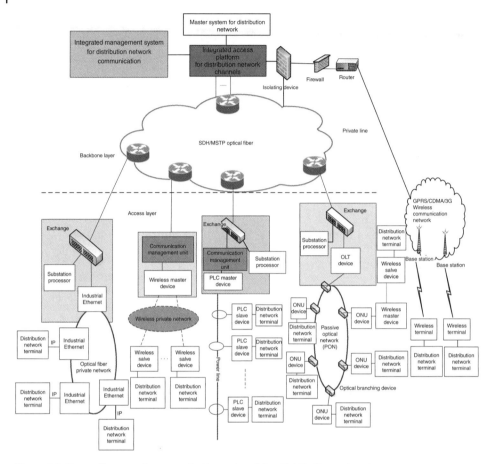

Figure 4.6 Frame diagram of communication network for distribution network.

3) *Access layer network.* The access layer communication network allows communication between master stations (substation) and terminal stations. At the time, adaptive protection for the regional distribution network is achieved via communication between substations and terminal stations. In so doing, the feeder line mentioned above performs feeder automation, automatically adjusts the protection settings, and automatically adapts to the mode of operation, independent of the dispatching layer or substation layer.

4) *Optical fiber private network (Ethernet passive optical network).* Communication between the substation and terminals uses Ethernet passive optical network EPON technology, which consists of OLT/ODN and ONU. ONU is configured at the terminal and connected to the terminal via an Ethernet interface or serial port. OLC is generally configured within the substation and integrates/incorporates the data into backbone layer communication.

5) *Optical fiber private network (industrial Ethernet).* When the industrial Ethernet is used as a means of communication, slave devices and terminals will be connected via an Ethernet interface. Master devices are generally located in the substation and

responsible for collecting data at all points in the Ethernet self-healing ring, and connecting them to the backbone layer network.

6) *Distribution wire carrier communication networking.* The carrier communication network is composed of one master network and multiple slave networks in accordance with DL/T 790.32. One master carrier with multiple slave carriers is able to form a logic carrier network where the master carrier accesses information in the backbone layer network via a communication supervisory machine. When multiple carriers are connected to the communication supervisory machine, the machine needs to have the functions of a serial server and a protocol to monitor the condition of the carrier on-line. Moreover, it supports conversion between automation protocols.

7) *Wireless private network.* Set up a wireless base station in the substation and access the information at terminals in the base station via a wireless private network; each network should be configured with a wireless communication module to communicate with the base station. Communication supervisory machines are arranged within the substation to access the base station information, transfer protocols, and then connect them into the backbone layer communication network.

8) *Wireless public network.* The terminal of the distribution network is configured with GPRS/CDMA/3G wireless communication modules and connected to a wireless public network. In some cases, the terminal of the distribution network can access wirelessly a communication integrated access platform, so as to communicate with the master station. The master station acquires information about the distribution terminal and has good knowledge of the operation modes of the distribution network. With this, the master station can provide information for line transference, load prediction, and topology analysis.

5

Distribution Network Communication Technology and Networking

5.1 Introduction to Distribution Communications

Owing to the large amount of distribution network equipment, wide geographical distribution, dispersed nodes, harsh operating environment, and uneven distribution, the communication scheme has more special requirements imposed on it. The modes of access to the distribution network can vary with the communication network coverage in different regions. The network conditions include fiber, power-line carrier, wireless and private networks, so unified management and unified interface specifications have long been studied. Besides the communication bandwidth, communication distance, real-time network, and other conventional functions and performances, the distribution communication network needs to have the following functions.

1) *Information region logic isolation.* There are four secondary power system security protection zones, as shown in Figure 5.1.
   - Safety zone 1: real-time control zone.
   - Safety zone II: non-control production zone.
   - Safety zone III: dispatching production management zone.
   - Safety zone IV: management information zone.
   There are physical separations between the safety zones, while there are logic isolations between the businesses in the same safety zone. That is to say, there is one independent physical network in a safety zone and one independent logic network for each business. Any data from the bottom of the electrical equipment, and overhead data generated in the transmission and calculation, are likely to be information in one or more than one of the four large safety zones. It is also possible that the data will become a message of one or more than one independent business. Therefore, the data source may pass through a physical isolation device or at least logical isolation equipment, and get to the corresponding business safety zone.
2) *Business performance guarantee.* Different types of business information have different requirements for the performance of information exchange, but no matter what kind of business, physical media, or communication model, it will have specific requirements for the communication time delay, error rate, etc. Distribution communication requires the normal communication of any business, and should not take a toll on the normal operation of other business communications. The normal operation of business communications refers to time delay, frame loss rate, and accuracy.

Self-healing Control Technology for Distribution Networks, First Edition. Xinxin Gu and Ning Jiang.
© 2017 China Electric Power Press. Published 2017 by John Wiley & Sons Singapore Pte. Ltd.

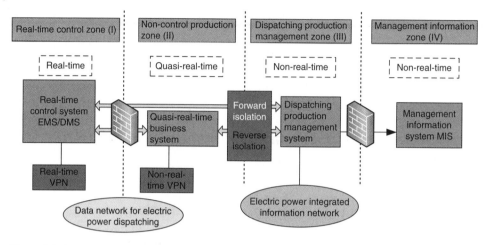

Figure 5.1 System security measures.

3) *Access security.* Since the network equipment has a wide range of types, the brands, interface types, and communication protocols are not the same. When various types of communication terminal are connected to a communication network of power distribution, all types of complex communication access environment need to be considered. We note the following two points.
   - Security measures for legitimate communication terminal access to distribution communication network: after legitimate communication terminal access to the communications network of power distribution, the correctness of the information transmission cannot be destroyed. What's more, its legitimate identity and physical equipment should be protected.
   - Precautions for illegal communication terminal access to distribution communication network: the distribution communication network may give an alarm and report it to the relevant departments or personnel once it senses illegal communication terminal access to the communication network. When the illegal communication terminal or disguised communication network attacks/damages the communication network of power distribution, the power distribution network can take preventive measures, minimizing damage to the network.
4) *Network administration system.* As the construction of the distribution grid develops, distribution network communication is in constant change. The capacity of resource allocation for the power distribution communication network should be improved as much as possible to keep the distribution communication network safe, rational, and efficient. The resources in the communication network of the power distribution include IP address, port, logical address in the communication protocol, and so on. To transmit information from the distribution equipment and sensors to the distribution network dispatching automation system, it needs to go through a layer of data gateways, and be connected to a communications network of distribution via a data gateway, to the backbone network via a substation, or with direct access to the backbone network and connected to the master station through a data gateway network, as shown in Figure 4.1. The data gateway can be one of FTU, DTU, or TTU. Geographically, the distribution substation, FTU, DTU, TTU, and power

distribution equipment are close to each other – within a radius of 10 km – while the distribution substation may be up to tens of kilometers away from the master station, or in a different city. Considering the transmission distance and the amount of information, technically, the backbone network and the power distribution communication network are different types of network.

5.2 Backbone Communication Network

5.2.1 SDH Technology

SDH (synchronous digital hierarchy), according to the suggested definition of ITU-T < baike. baidu.com/view/25538.htm>, is an information structure at a level corresponding to the transmission rates of the digital signal < baike.baidu.com/view/988371.htm>, including multiplexing and mapping methods, and a technology system composed of all synchronization methods. Hence, SDH is a practical solution to the application of a backbone communication network.

The SDH is integrated with a multiple connection, transmission lines, and switching functions, and under the management of a unified gateway. In this way, it can support packet switching (four-remote control system), streaming (video surveillance) in a private communication protocol of a power distribution business, and other demands for information transmission. Compared with PDH (plesiochronous digital hierarchy), SDH has the following advantages.

1) *Better standardization.* Standardization is represented as standard interface and bit rate. The requirement for standards simplifies interconnecting work between equipment of different brands and types.
2) *Better ability of channel multiplexing.* An SDN network is like a "high-speed synchronous railway" with information from all kinds of device. Communication protocols are like the goods that are delivered by the high-speed synchronous railway. In an SDN network, the goods to be delivered are loaded into compartments of the train in a high-speed synchronous railway by way of byte interleave, and all goods are indexed for ease of distribution after unloading. Besides, the SDH network guarantees that goods are shipped to their destination on time and in good condition.
3) *Better network self-healing ability.* An SDH network device supports a looped network through which it is able to look for other possible transmission channels and realize recovery communication in case the main signals are interrupted. At the same time, it uploads the transmission-medium fault signal to the power gateway system.
4) *Better ability of operation management.* An SDH network frame structure is arranged with overhead byte having a complete operation management maintenance function, making the OAM ability of the SDH much higher than that of a PDH network, effectively reducing overall maintenance costs.

5.2.2 MSTP Technology

MSTP (multi-service transfer platform, based on the SDH multi-service transfer platform) refers to SDH platform-based multi-service nodes that have the ability to access, process, and transport TDM, ATM, and Ethernet services and provide a unified gateway.

In the field of distribution communication networks, EOS (Ethernet over SDH) technology is most commonly used in MSTP. With the rise of the MAN integrated service market, EOS continues to improve. Currently, the popular data-encapsulating mapping protocols are PPP/ML-PPP, LAPS, and GFP. There are no strict limits on Ethernet-to-VC encapsulation in the MSTP standard, therefore, any of the mainstream protocols can be flexibly applied to equipment of any brand.

Figure 5.2 is a business process model of the MTSP technical specification stipulated. Modules within the red dotted line are a series of process actions before SDH mapping. The green outlined/shaded boxes show three protocols that are all able to encapsulate Ethernet data, and each of them has its own technical background and characteristics.

5.3 Distribution Communication Technology

This section describes network technologies similar to the distribution equipment, except for the backbone communication network, including a passive network, industrial Ethernet, wireless communication, and power-line carrier. These technologies have advantages in different circumstances, and will be briefly described one by one.

5.3.1 EPON

A PON (passive optical network) refers to an ODN (optical distribution network) that doesn't contain any electronic components or electric power supply; the ODN is composed of a splitter and other passive devices, instead of expensive active electronic devices. One PON is composed of an OLT (optical line terminal) at center station and matched ONUs (optical network units) at the consumer's site. The ODN between the OLT and ONUs includes an optical and passive splitter or coupler.

An EPON (ether passive optical network), just as its name implies, provides a wide range of services in the Ethernet by point-to-multipoint and passive optical transmission based on Ethernet PON technology. As the mainstream optical access technology in distribution network communication, EPON has distinct advantages in anti-interference, bandwidth, access distance, and maintenance management.

5.3.1.1 EPON Technology and Characteristics

The EPON structure is composed of an OLT, ODN, ONU/ONT, and EMS (element management system) in a point-to-multi-point tree topology. The EPON network is able to supply power to outdoor network devices, and signal forwarding/switching will be completed within the switch and user equipment, greatly reducing the cost of network channels. With less front-end capital investment in network deployment, more money can be put into user terminal devices. The probability that faults will occur in an outdoor EPON is much lower (because many faults occurring in an active network result from power supply problems). Using EPON, a network can transmit a distance of about 20 m (less than an active optical access system). It also has a shorter-range coverage, but meets all the requirements when used for distribution communication (a 10-kV line has a transmission distance of 10 km).

As shown in Figure 5.3, an EPON network provides a wide range of services above the Ethernet by taking advantage of the tree topology and passive optical transmission. It

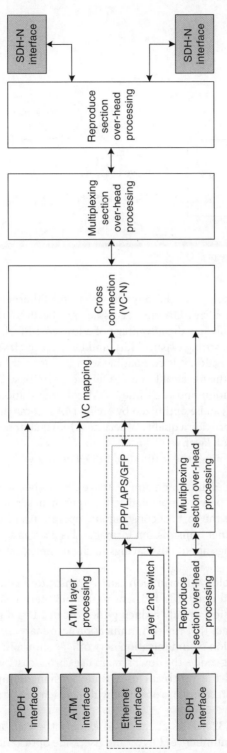

Figure 5.2 MTSP business process model as technical specification stipulated.

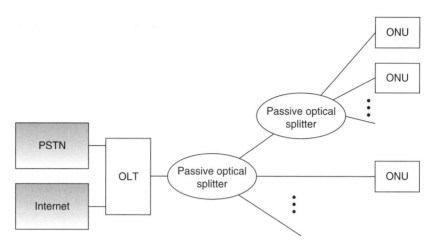

Figure 5.3 Network structure of EPON.

also applies PON technology to the physical layer and Ethernet protocol to the link layer, and successfully realizes Ethernet access using the PON topological structure. Therefore, EPON not only has the advantage of a low-cost PON, but also has the features of high bandwidth, strong expansibility, and flexible/quick service reconstructing that the Ethernet has. Besides, it is compatible with the present Ethernet and convenient in management. At the physical layer, EPON uses a 100Base Ethernet PHY chip. If abrupt data communication and real-time TDM communication take place between the ONU and OLT, this can be optimized by giving MAC commands at the PON layer. In addition, EPON will provide a quality of service (QoS) guarantee similar to APON by virtue of 802.1p at the link layer.

This framework determines the features of EPON as follows:

1) There are only optical, splitter, and other passive devices between the OLT and ONU, so the machine room, power supply, and active equipment servicemen are not needed, which greatly reduces the costs of construction, operation, and maintenance.
2) EPON uses the Ethernet transmission format and is used as a popular technology in LAN/CPN. This brings about cost savings in the complex transformation of transmission protocols.
3) Using SWDM technology, information can be transmitted as far as 20 km with only one trunk optical and one OLT, and sent to up to 32 terminals through an optical splitter at the ONU side, greatly reducing the cost of LT and trunk optical. This is consistent with the fact that the communication terminals are locally close to each other in the distribution communication network, as shown in Figure 5.4.
4) In full duplex working mode, the downstream uses broadcast encryption suitable for various services and the upstream uses time division multiple access (TDMA). Upstream and downstream share one bandwidth, which can be adjusted for different businesses.
5) One OLT can carry 32 ONUs at most. To expand the capability or add terminal users, just add small numbers of ONUs and user-side optical fiber.
6) With enough bandwidth, the upstream rate on each ONU reaches at least 800 Mb.

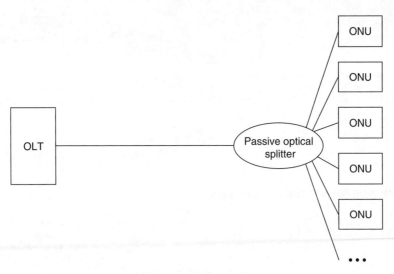

Figure 5.4 Signal transmission of EPON system.

7) The support TDM, IP data, and data transmission of videocast, and performances, meet the requirements of distribution communication.
8) Low cost, short construction period, easy to spread.
9) Better QoS.

5.3.1.2 EPON Interface

EPON is backbone network-oriented upwards, with OLT as boundary interface. Some technologies are pioneers in trying a 10Gbit/s interface, mainly the GE (Gigabit Ethernet) interface or other interfaces according to need. This is orientated to DTU, FTU, and TTU devices downwards, with ONU as boundary interface, and equipped with a passive optical interface.

As a boundary device connected to a backbone network, besides switching and routing, OLT also supports an SDH interface standard at rates of ATM, FR, and OC3/12/48/192. OLT can expand interfaces when it is configured with a multiple OLC (optical line card). OLC is connected to ONUs through passive splitters. Generally, the splitting ratios of POS are respectively 8, 16, 32, and 64. Multi-level connection can be achieved, since the maximum number of ONUs that LT and PON can be connected to is related to the equipment/devices. Besides, OLT supports an E1 interface and realizes traditional TDM video access. In terms of EPON unified network management, OLT is the most important control center to realize web-based management.

5.3.1.3 EPON Transmission System

WEPON realizes transmission between uplink and downlink through a single optical by use of WDM (wavelength division multiplexing). The bandwidths for uplink and downlink are symmetrical, both being 1.25Gb/s.

At the downlink, since every OLT corresponds to more than one ONU, each ONU will broadcast an allowable time slot for access to that ONU on a regular basis. The ONU will initially make a login request upon receipt of a broadcast. All ONUs are

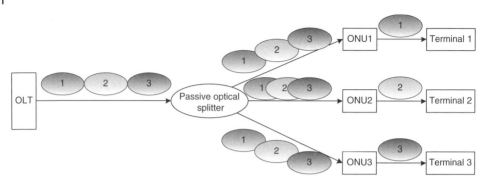

Figure 5.5 EPON downlink transmitting principle.

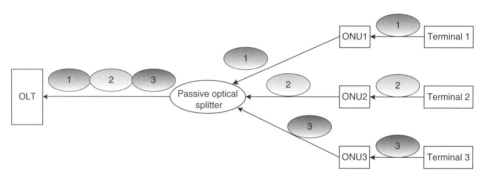

Figure 5.6 EPON upstream transmission principle.

under authentication management, and a unique LLID (logical link identifier) is allocated to each ONU. When the OLT sends a message to ONU1, 2, and 3, as shown in Figure 5.5, the OLT will broadcast a message outside. The agreed positions for each piece of the message include the LLID of ONU1 and 2. An optical splitter broadcasts the message down to every ONU. When ONU1 receives package 1, 2, and 3, it sends only package 1 to terminal user 1, dropping package 2 and 3. This is also true of ONU2 and 3.

At the uplink, each ONU sends information to its own OLT alone, and the message can't be transmitted directly among ONUs. TDMA is used for uplink transmission, and the LLDI has a mapping relationship with the time slot of OLT management, since the LLID is allocated to each ONU by the OL. Every ONU writes the agreed position of the uplink message into the LLID and then sends the message to a splitter, which arranges the message by order of slot time and sends it to the OLT, as shown in Figure 5.6.

5.3.2 Industrial Ethernet

All Ethernet devices are industrial equipment. An industrial Ethernet switch and router have similar functions to a commercial Ethernet switch and router, but they can meet the needs at sites in terms of product design, component selection, environmental requirements, redundant power supply, installation, thermal design, housing design, protection class, safety, and reliability.

An industrial Ethernet switch is mainly used for real-time Ethernet data transmission in complex industrial environments. Since the introduction of a carrier sense multiple access/collision detection (CSMA/CD) system in complex industrial environments, the reliability has been greatly reduced, meaning that ordinary Ethernet equipment and materials cannot be used. The use of storage conversion exchange in an industrial Ethernet switch increases the rate of Ethernet communication. What's more, the built-in smart alarm is designed to monitor network operations, ensuring that the Ethernet is operating reliably and stably in harsh and hazardous industrial environments.

The parameters of an industrial Ethernet switch and general commercial Ethernet switch are displayed for contrast in Table 5.1.

By contrast, an industrial Ethernet switch has the following characteristics:

1) components and parts of the product are industrialized;
2) adapted to harsh environment in industrial fields;
3) redundant power supply and wide voltage;
4) mechanical structure and method of installation;
5) heat dissipation of fanless product;
6) high-strength alloy shell;
7) IP grade structure;
8) redundant network link failure recovery time < 50 ms.

5.3.3 Wireless Communication

There is a wide variety of wireless communication technologies, with many grouping methods by different criteria. For example, they can be classified into single, half-duplex, or duplex systems by working mode; they can be classified into FDMA, TDMA, or CDMA by multi-access method; they can be classified into satellite communication

Table 5.1 Industrial and commercial Ethernet switch parameters

Item	Industrial environment	Office environment
Working voltage	24 V DC/48 V DC/110 V DC/220 V DC	220 V AC
Working temperature	−40 °C to +85 °C	0 °C to +50 °C
Installation	Card track/rack	Rack
Shock-resistant impact index	2 g/15 g	None
Heat-dissipating method	Fanless, shell	Fan
IP grade	IP30 ~ IP67	IP20
Redundancy requirements	Combination of redundancy modes < 100 ms	No or core redundancy < 1 s
Standard	Industrial EMC standard	General standard
Operating life	>10 years	>3 years
Harsh industrial environment	Strong electromagnetic environment/ intrinsically safe/explosive proof	Office environment
Corrosion protection	Three proofing	No protection measure

systems, short-wave communication systems, or scatter communication systems by transmission means; they can be classified into short-distance, middle-distance, or long-distance wireless communication technologies by distance of wireless communication.

Short-distance wireless communication technology refers to technology that has a short communication distance, covering about 10–200 m (e.g., Bluetooth, UWB radio, Wi-Fi, ZigBee, etc.). Middle-distance wireless communication technology refers to technology that has a communication distance of ten to several hundred kilometers (e.g., microwave and 230 MHz data transfer radio); long-distance wireless communication technology refers to technology that covers several hundred kilometers or above (e.g., GPRS, 3G, and LTE).

Currently, the most commonly used technologies in distribution communication are the following.

1) *GPRS, 3G, LTE.* Owing to limitations of geographical environment, high construction costs, or dispersed physical positions between communicating nodes (dispersed here means a distance of 400 m or more), it is difficult to realize full coverage of a wired network in power distribution areas. In such circumstances, the power distribution communication nodes are connected to a specialized web of telecommunication operators by use of a wireless communication module, and a wired data gateway is deployed in places where there is a good-quality Internet service (later accessing the backbone network). In this way, the requirements are met. The technology is limited by the transmission rate between upper and lower communication nodes, since too much exchange of information between upstream and downstream nodes will lead to high costs of traffic communication flow. Generally, the use of wireless communication flow is controlled by limiting the period of upstream and downstream information exchange, since information is counted per month. So, the technology is applicable for small amounts of enduring, high-delay information.

2) *Short-distance communication.* Wi-Fi, Bluetooth, and ZigBee are all applied in short-distance communication. Owing to the limitations of the environment, some equipment can't be used for laying cables, although it is equipped with a communication interface. In this case, it can hardly realize informatization of the distribution equipment. When adding devices or replacing existing equipment with the addition of cables, large-scale power-off maintenance is necessary, so it is not operational in practice. Establishing a LAN for power distribution devices that are distributed densely at a physical location by virtue of short-distance communication technology and then connecting it to a distribution communication Internet (like an EPON) via a LAN gateway will reduce the construction costs and make it more convenient, providing a more flexible scheme for informatization of electrical equipment.

5.3.4 Power-Line Carrier

The power-line carrier is a kind of communication technology which uses the power line as transmission medium. It is most suitable for the power industry, since the spatial position of an informatized physical channel of the power equipment is similar to that of a physical channel of power energy transfer. However, due to the limitations of stability, bandwidth, and supported service transmit type, the power-line carrier is no longer the mainstream technology used in power communication, and is only supplemented/reserved for the optical communication network.

The power-line carrier communication transmits the processed information from the sending terminal to the power line; then, the power line transmits the information to the opposite terminal; finally, the information is separated from the power energy through processing and transferred to the receiving terminal. The power carrier communication is composed of a signal processing device and a signal coupling device. Its information transmission procedure is shown in Figure 5.7. First, the information enters a signal processing device through the data interface; then, the signal processing device carries out data coding, channel coding, and channel modulation for such information and sends the modulated signal to the coupling device; next, the coupling device couples the carrier communication signal to the power line and transmits it to the opposite terminal; then, the coupling device of the opposite terminal separates the carrier signal from the power line; finally, the signal processing device restores the original data through channel demodulation, channel decoding, and data decoding. The data transmission is subject to half-duplex and totally restrained by the communication protocol of the application layer.

The performance indicators of the power-line carrier are as follows.

1) The networking structure of carrier communication is a master–slave tree-form topology network.

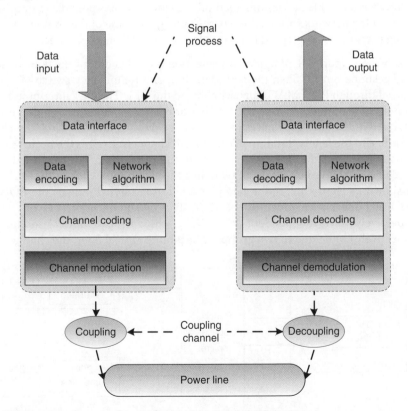

Figure 5.7 Structure diagram of power-line carrier communication system.

2) The frequency band range of the carrier communication system is approximately 20–500 kHz.
3) The signal power requirement of the carrier communication system is < 5 W.
4) The receiving sensitivity of the carrier communication system is < 1 mV.
5) The communication error rate of the carrier communication system is < 10^{-6}.
6) The nominal output impedance of the carrier communication system is 75 Ω.
7) The physical interface of the carrier communication system is an RS232, RS485, and Ethernet interface.
8) The communication protocol of the carrier communication system is 101/104/CDT and other power system protocols.
9) The power consumption requirements of the carrier communication system are static power consumption < 5 W; instantaneous power consumption of sending < 20 W.
10) The power supply requirements of the carrier communication system are 48 V DC and 220 V AC.

5.4 Communication Networking Method of Power Distribution

5.4.1 Basic Topology

Since EPON is a flexible networking method with low cost, it is possible to improve the reliability of the network through hot spare/topology of some of the equipment. In the following paragraphs, the typical basic topology form is taken as example.

1) As shown in Figure 5.8, redundancy protection is carried out for the optical module and backbone optical fiber, the fault detection and redundancy switch of which is realized through PON MAC chip switching in the OLT. The OLT interior judges the running state of the backbone optical fiber or optical module through the real-time detection of the backbone optical fiber and decides whether to switch the channel according to the failure duration of the current channel as well as the occurrence and switching time of the last communication failure of the other channel. Channel switching is achieved by means of switching the electronic switch in the OLT.
2) As shown in Figure 5.9, redundancy protection is carried out for the PON port and backbone optical fiber of the OLT. The two PON interfaces of the OLT respectively have independent PON MAC chip; its two optical modules are respectively connected to two groups of backbone optical fiber; the channel is still

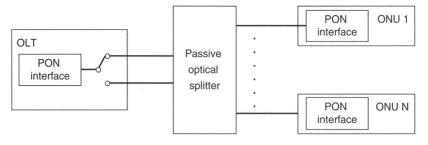

Figure 5.8 Redundancy protection mode of backbone optical fiber.

judged by the OLT; then, channel switching is realized through enabling different PON ports.

3) As shown in Figure 5.10, redundancy protection is carried out for the PON interface, backbone optical fiber, optical splitter, distribution optical fiber, and ONU optical module. In addition to the OLT, the PON interface of the ONU is also able to judge the channel failure and switch channels. The OLT is still subject to active switching, while the ONU is subject to passive switching. In this way, the consistency of master and slave channels can be ensured among all ONUs.

4) As shown in Figure 5.11, redundancy protection is carried out for the OLT PON interface, backbone optical fiber, optical splitter, distribution optical fiber, and ONU PON module. This mode is the top redundancy configuration in the EPON tree-form topology. The damage of any OLT PON interface, backbone optical fiber,

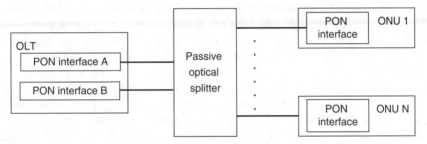

Figure 5.9 Redundancy protection mode of OLT PON.

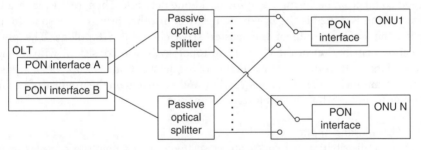

Figure 5.10 Redundancy protection mode of entire optical fiber.

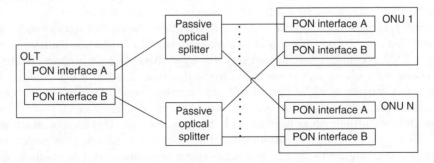

Figure 5.11 Redundancy protection mode of OLT PON interface.

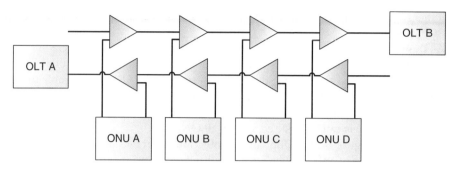

Figure 5.12 Hand-in-hand protection mode.

optical splitter, distribution optical fiber, or ONU PON interface will not result in service interruption or information loss. However, suppose we divide all sections into two corresponding groups A and B. If some section of group A faults and the fault is not removed in a timely fashion, and some section of group B also faults, then normal service will be affected and interrupted. Therefore, it is still necessary to remove the fault in any network of redundancy configuration in a timely manner. Otherwise, further expansion of the fault will influence normal service eventually.

5) Figure 5.12 presents a special topological form. It is not a simple tree structure. We call it a 'hand-in-hand' topology. Two chain-form topologies, composed of OLT, backbone optical fiber, and optical splitter, are connected to each ONU through a branch of the optical splitter. Two OLTs are distributed in different locations, which are usually far apart from each other. However, they are arranged in positions with good connection performance to the backbone network. Then, real-time intercommunication is done via the backbone network or other private line, so as to determine the working state of the master–slave chain-form topology. The physical location distribution of the chain-form topology is also consistent with the actual conditions of the transmission structure of the distribution network. So, it is possible to carry out maintenance, reconstruction, and extension on the premise of having no influence on the whole informatization function.

5.4.1.1 Networking Application
Although the distribution network features scattered access points, a complex environment, and continuous construction, the EPON still applies to the distribution network based on its low cost, good bandwidth, flexible networking, and good service compatibility. It is the preferred solution for a distribution communication network.

Figure 5.13 presents a typical case of a distribution communication network composed of an EPON. The OLT and following parts form the distribution communication network; OLTs compose the ring network, the upper part of which is connected to the backbone network. Here, the backbone network is simplified into two routers. The lower part extends out four optical splitters, forming three informatization areas of the distribution network: areas A, B, and C. Area A is extended from one OLT, which composes the tree-form topology together with the optical splitter. This form is usually used for comparatively standard substations, featuring large amounts of information, centralized equipment, and a clear layer. Area B is extended from two OLTs, composing the hand-in-hand topology together with the optical splitter. This form is usually used for

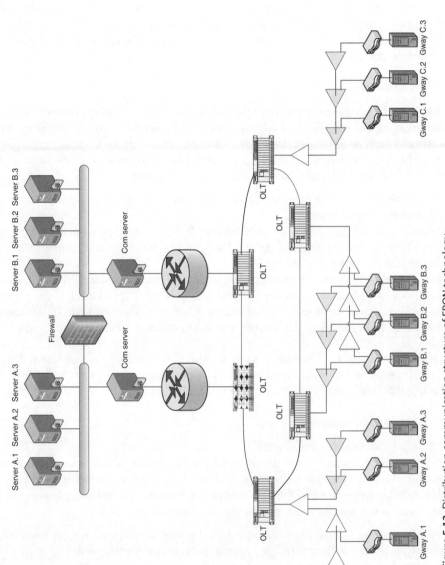

Figure 5.13 Distribution communication structure of EPON technology.

informatization of distribution-line equipment and extends with the distribution line, which has sufficient reliability. Area C is extended from one OLT, which composes the chain-form topology together with the optical splitter. This form is usually used for scenes featuring low requirements for stability and real-time performance, and scattered node locations (e.g., concentrated meter readings).

Compared with SDH, EPON has a lower cost. It applies to distribution networks of huge scale and greatly reduces the power-supply and maintenance problems of outdoor equipment. Therefore, EPON has been promoted successfully and used extensively in distribution communication networks.

5.4.2 Industrial Ethernet

The greatest advantages of industrial Ethernet include: its technology is mature; customers are comparatively familiar with this technology; its function and performance indicator can both meet the demands of distribution communication. However, its cost is slightly higher than that of EPON. So, it is difficult to deploy in some cases, or cannot be realized. The industrial Ethernet has the following features.

1) *Flexible networking.* It is capable of realizing chain-form, tree-form, star-form, and ring-form or hybrid topology.
2) *Dual-network hot standby.* Two networks with the same physical equipment are used to ensure communication quality. Network use depends on the equipment in the application system and the network.
3) *Power-supply redundancy.* The switch, router, optical transmitter, and receiver all use a DC power supply. In many cases of distribution network, the condition of the power supply is very poor and the frequency of power failure is relatively high. In this case, power-supply redundancy can reduce the communication fault caused by power damage.
4) *Network management.* The functions of VLAN, DHCP, priority management, flow control, and network storm suppression ensure communication quality and safety of the IP data stream, including MMS, GOOSE, and other protocols.

5.4.3 Wireless Communication

5.4.3.1 Short-Distance Communication

Short-distance communication is usually used for data transmission from a sensor or some controllers. Its greatest advantages include convenient construction and more flexible installation position of the data source equipment. The following features are present in most cases using short-distance wireless communication.

1) *Low power consumption.* No need for external power supply, with battery installed. Able to charge the battery through a mutual inductor or electric field.
2) *Low rate.* Usually, it is not convenient to lay the communication cable around high-voltage equipment or on telegraph poles. In such cases, less data needs to be transmitted, most of which refers to two-remote signals measured by sensors. The requirement for transmission delay is also not high.
3) *Low cost.* The closer the data is to the source, and the more scattered the data, the more data there is. If there is no cost advantage, the life force of the technology/product will become vulnerable.

Table 5.2 Feature comparison of Bluetooth, Wi-Fi, and ZigBee

	Bluetooth	Wi-Fi	ZigBee
Frequency band used	2.4 GHz		
Price	Moderate	Expensive	Cheap
Coverage area	Theoretical area: 100 m Actual area: about 15 m	100 m	75 m
Power consumption	Low	High	Low
Transmission speed	1 M/s	54 M/s	250 k/s
Equipment connectivity	7	50	50
Safety	High	Low	High

4) *Transmission distance.* About 50–100 m. Offside wireless communication equipment should be deployed within this range.

From Table 5.2 it can be seen that for Bluetooth, Wi-Fi, and ZigBee, the ZigBee technology can best meet the requirements for the above cases of distribution communication, which need short-distance wireless communication. In case 2.4 GHz is used in all such technology, although ZigBee has slightly less coverage area and low transmission speed, it features low cost, low power consumption, and satisfactory transmission speed. Therefore, ZigBee is a comparatively more suitable short-distance wireless communication technology for distribution networks.

5.4.3.2 TD-LTE

TD-LTE (time division long-term evolution) is jointly formulated by global enterprises and service providers covered by the 3GPP organization. The base station of TD-LTE includes BBU (baseband unit) and RRU (remote radio unit). The BBU is usually arranged in the high-voltage substation of the distribution network. The RRU and RF antenna are usually installed in a distribution area transformer, RMU, and pole-mounted switch, which connect with the BBU via optical fiber. At most six RRUs can be installed under one BBU. Besides, each RRU can be used as an optical fiber repeater and subject to a cascading connection. The core network above the base station is usually arranged in the control center of the urban power bureau. Service terminals, such as wireless terminals, connection concentrators, and distribution network equipment, are connected with each other through a wired mode.

In Figure 5.14, a typical example of TD-LTE network deployment is presented. It is connected to the substation in the data center of the urban power bureau through the backbone network; a set of BBUs is arranged in the substation; then, two circuits of optical fiber are pulled out from the BBU. One circuit of optical fiber is connected to RRU1 and extends downwards through RRU1 cascading, reaching RRU2 or even further. The other circuit is connected to the other branch, covering other areas. One side of the wireless terminal communicates with RRU3 through an RF signal and the other side with the distribution equipment through an Ethernet or serial port.

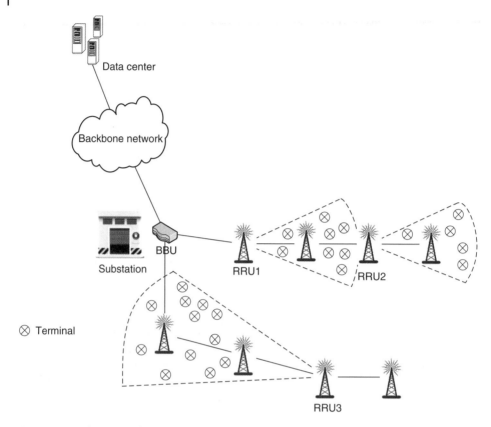

Figure 5.14 Architecture of TD-LTE network deployment.

5.4.4 Hybrid Networking

The typical programs of EPON, industrial Ethernet, and wireless communication technology used for distribution information networking have been introduced in the above paragraphs. However, in practical applications, the entire distribution communication network is structured through a combination of many kinds of technology in most cases. These technologies can be divided into optical fiber, power-line carrier, and wireless according to the transmission medium. Both EPON and industrial Ethernet are optical fiber communication technology. Since these three kinds of technology have different economic efficiency, real-time performance, reliability, safety, and stability, it is necessary to select the corresponding technology according to the type of access service, concrete application scene, and construction investment budge when constructing the distribution communication network. Table 5.3 presents the respective advantages and disadvantages of the three kinds of communication technology.

At present, the basic construction plan of a distribution communication network is as follows: focus on optical fiber communication; supplement with a power-line carrier and wireless communication; form a networking solution through mixing of multiple kinds of technology.

Table 5.3 Comparison of communication technologies

Communication technology	Advantages	Disadvantages
Optical fiber	• High bandwidth, easy networking, and convenient access. • High reliability, high real-time performance, and high safety. • Little interference from ambient.	• Comparatively high construction cost, mainly refers to the cost of optical cable.
Carrier wave	• Low construction cost. • Convenient construction with no need for wired arrangement. • Operation in private network and high safety.	• Low bandwidth; • Low reliability and real-time performance; • Poor network flexibility;
Wireless	• Low construction cost. • Convenient construction with no need for wired arrangement. • Wide application range and high availability.	• Low real-time performance and safety. • Easily disturbed by ambient. • Restricted frequency band and power.

The general framework of distribution communication in Figure 5.15 integrates optical fiber technology, power-line carrier technology, and wireless communication technology. Optical fiber communication is mainly used to construct the backbone network, industrial Ethernet, and EPON network; the power-line carrier is usually connected to the backbone network via chain-form topology through the DTU; the short-distance wireless communication locally forms a local area network and is then connected to the optical fiber through the wireless master station and on to the backbone network. Besides, in some cases, short-distance wireless communication can also be connected to the backbone network through a TD-LTE private network of telecom operators.

Since the optical fiber, power-line carrier, and wireless communication can connect with each other through a serial port or LAN port, there is not much difficulty in theory in integrating many kinds of communication technology. The networking of many kinds of communication technology falls into supplementary form and standby form. Based on the present guiding thoughts on optical coverage first, areas where the optical fiber cannot be laid need to be remedied through power-line carrier or wireless communication, which is called the supplementary form. When it is possible to cover some target area through two or more kinds of communication technology, many different kinds of communication technology are used to establish several channels to improve the communication reliability. In this way, when one channel is working, other channels are in a state of cold/hot standby. This is called the standby form.

The standby form is constructed completely in accordance with unrelated independent channels with no specialty in technology. The channel is evaluated and selected via the application software at the time of use. The judgment and management of the channel master–slave state from the software is not introduced much here. In the following paragraphs, the supplementary form is mainly introduced, including optical fiber + power-line carrier, optical fiber + wireless, and power-line carrier + wireless.

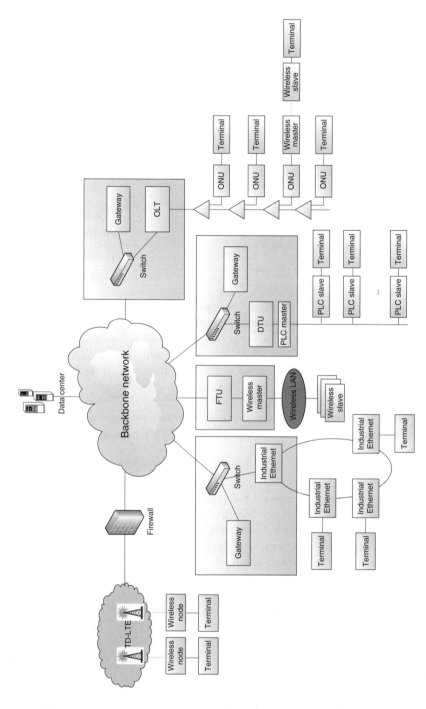

Figure 5.15 General framework of hybrid networking through multiple kinds of communication technology.

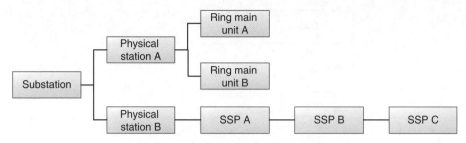

Figure 5.16 Primary wiring diagram of distribution network.

5.4.4.1 Optical Fiber + Power-Line Carrier

As shown in Figure 5.16, the area from substation to both civil engineering stations A and B can be networked through an EPON, and optical fiber can be laid in such an area. However, optical fiber cannot be laid in the area from civil engineering station A to RMUs A and B, and from civil engineering station B to switching stations A, B, and C. According to the principle of optical fiber first and power-line carrier supplementing, a master station of the power-line carrier is respectively deployed in civil engineering stations A and B. Then, a slave station of the power-line carrier is respectively arranged in RMUs A and B, which forms a set of carrier local area network together with the master station of the power-line carrier of civil engineering station A. Besides, a slave station of the power-line carrier is respectively arranged in switching stations A, B, and C, which forms a set of carrier local area network together with the master station of the power-line carrier of civil engineering station B. These two carrier local area networks are led out through the two master stations of the carrier of civil engineering stations A and B and connected to the ONU of the EPON network (through a serial port or industrial Ethernet), totally realizing the network coverage of two civil engineering stations, two RMUs, and three switching stations.

5.4.4.2 Optical Fiber + Wireless

As shown in Figure 5.17, the backbone network connects with an industrial Ethernet, which then leads out two branches. One branch is connected to a wireless gateway, which covers some wireless hotspots in the formulation area through an RF pole and tower. Then, the wireless hotspot forwards the information from the connected distribution equipment to the distribution communication network through wireless. The other branch is connected to a group of an EPON network. First, the industrial Ethernet switch leads out the optical fiber to the OLT; then, some ONUs are installed downwards through an OLT and an optical splitter; the ONUs are downward connected with the wireless communication data gateway, which then maintains the wireless local area network of the lower layer, thus realizing a combination of industrial Ethernet, EPON, and short-distance wireless.

5.4.4.3 Power-Line Carrier + Wireless

The power-line carrier is capable of bearing the service with a bandwidth within the range of 2–20 M and can be used as the uploading channel of several kinds of terminal station service. Meanwhile, based on the long-distance transmission capability of a power-line carrier, the data collected by wireless concentrator can be uploaded via

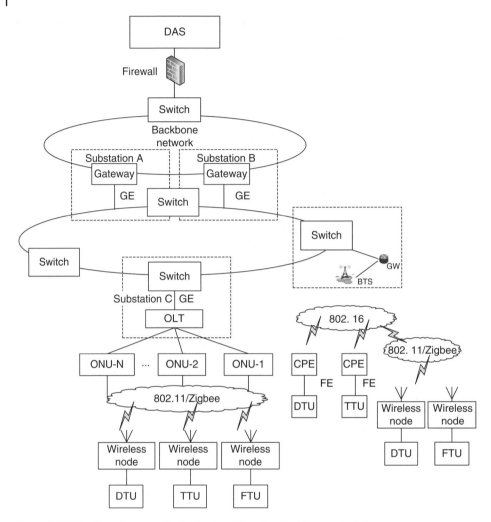

Figure 5.17 Structure diagram of hybrid networking of optical fiber and wireless.

a power-line carrier. The master carrier of the power-line carrier can not only be connected directly to the master communication station, but also reach the master communication station through other communication means (e.g., broadband wireless network or optical network).

The short-distance wireless ZigBee technology can be used as terminal data, which can be collected and transmitted to the convergence station (TD-LTE, 230 MHz, and so on). Since the ZigBee technology is featured with short transmission distance, the ZigBee network service is usually converged with the broadband wireless or optical fiber channel for uploading.

The power broadband wireless communication system (e.g., TD-LTE, 230 MHz, and so on) can realize 3–5 km transmission and 1–20 M data bandwidth, meeting the demands of most services, which can not only substitute the optical fiber network to bear the access communication of the last kilometer, but also be used as the master

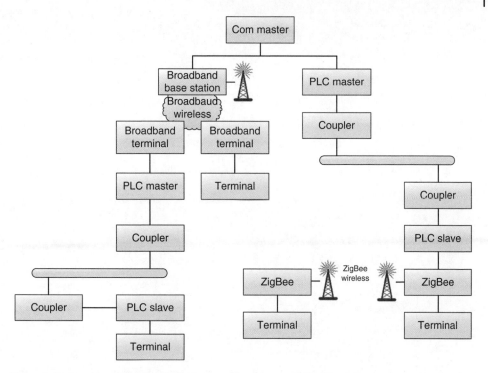

Figure 5.18 Structure diagram of hybrid networking of many kinds of communication network.

channel to upload the service. Compared with the power-line carrier, the wireless communication system has the higher construction cost. Therefore, it has to be considered and used according to the on-site application environment.

The hybrid networking of many kinds of communication network, as shown in Figure 5.18, includes power-line carrier technology, short-distance wireless, and broadband wireless. The medium- and short-distance wireless is used as terminal data access equipment and service is converged with the power-line carrier and broadband wireless to make up the short transmission distance of ZigBee. The power-line carrier technology can be used not only as terminal data access equipment, but also as wireless communication convergence equipment to finish the service uploading function of the terminal or wireless communication equipment, thus remedying the problems of optical fiber not being laid and blind areas of wireless communication caused by buildings or other factors. Broadband wireless communication can play the same role with a power-line carrier, and be used as emergency communication to play a part in equipment maintenance.

This kind of networking form based on a power wireless broadband communication system and power-line carrier short-distance wireless technology can be applied in urban distribution networks, thus efficiently reducing the investment required. Furthermore, this program can also be used as an emergency communication system, used to meet the special communication requirements in case optical communication is interrupted when the optical fiber channel acts as convergence station. The interfaces of all communication equipment have to meet some requirements in the hybrid networking – i.e., they should be able to support each other's working mode.

6

Detection Management System for Distribution Network Devices

6.1 Significance of Distribution Equipment Condition-Based Monitoring and Maintenance

The distribution network is electricity consumer–oriented and undertakes the task of distribution of electricity locally or by voltage levels, therefore it is one key link to control or ensure power supply. However, for a less-developed distribution network, it is not practical to configure most distribution lines with reference to a design principle of transmission line $N-1$. At present, above 95% of power outages are caused by failures in the distribution system. So, whether the distribution system and distribution equipment are reliable makes a great deal of difference to the reliability of power supply.

There is a significant difference between power distribution and transmission/transformation in terms of operation and maintenance. Comparatively speaking, a decentralized layout and the wide variety of types/structures that characterize a distribution device make maintenance work more complex. Taking equipment failures, for example, according to statistics of the SGCC, 110 kV and below switchgears represented more than 70% of all equipment failures in the year 2004. Other contributory factors included: large amount of distribution equipment; relatively low prices; strict cost control on operation and maintenance. Although the reliability of a distribution network is improved by assigning more maintenance personnel on site and carrying out maintenance on a more frequent basis, this is not an economical or practical solution in the long term. Traditional time-based preventive maintenance may reduce the maintenance costs to some degree, by simply extending the maintenance intervals, while it has the potential risk of greatly increasing primary equipment failure. Once failures are found, the increased costs arising from maintenance and replacement may be much higher than the costs saved by extending the maintenance periods. Besides the ongoing efforts to improve power supply services and optimize operation patterns, power grid enterprises should make a careful analysis of distribution equipment characteristics during operation and maintenance. For this reason, it is necessary to introduce condition-based maintenance as a means to improve the operation and maintenance level after experiential learning from power generation, transmission, and transformation.

Condition-based maintenance is a scheme of maintenance where the maintenance items and times are determined depending on the results of equipment condition evaluation, risk assessment, and fault diagnosis based on equipment condition evaluations

Self-healing Control Technology for Distribution Networks, First Edition. Xinxin Gu and Ning Jiang.
© 2017 China Electric Power Press. Published 2017 by John Wiley & Sons Singapore Pte. Ltd.

(including condition monitoring, life prediction, and reliability-centered assessment); proactive maintenance is carried out to ensure that operations are safe and reliable, and costs are reasonable.

The central task for condition-based maintenance is to determine the condition of equipment according to which relevant tests and inspections are going to be developed. As mentioned above, condition-based maintenance does not simply involve extending the maintenance periods or reducing the maintenance costs; instead, the focus is shifted from maintenance or repair to management. Condition-based maintenance uses all means of monitoring, testing, and maintaining to obtain the required characteristic parameters and make an accurate evaluation of the conditions and reliability of equipment based on the effective acquisition, organization, and analysis of data until a scientific scheme of maintenance is developed by operation managers to find abnormalities in a timely and effective manner, so as to reduce the test accompanying rate and maintenance accompanying rates under the original pattern. It can be seen that in order to carry out condition-based maintenance, it is necessary to enhance condition monitoring and information management and keep them under strict control via a breakdown analysis using information management and technology analysis. We strive to do the best.

The goal of condition-based maintenance is to improve the continuity of distribution equipment by optimizing and implementing an equipment maintenance schedule based on the results of reliability-centered monitoring and supervision, and further improve the operation and business management level. As we can see, condition-based maintenance does not simply involve reducing the maintenance workload or extending the test periods, but rather reducing the test accompanying rate and the maintenance accompanying rate in the equipment condition evaluation in a standard, lean, and scientific management approach. For defective and dangerous equipment, the monitoring and maintenance or overhaul should be enhanced to eliminate faults as soon as possible, so as to make maintenance more targeted and effective and improve the efficiency of the equipment as well.

To meet the demand for development of distribution networks and distribution management systems (DMSs), it is necessary to integrate data about conditions into the system. It is true that SCADA, AM, FM, and GIS in the distribution network will be able to improve electric power distribution and power dispatching, routine maintenance, and urgent repair in case of emergencies. However, with increasing capacity and class, there is an increasing amount and category of electric equipment to be monitored and supervised, which is more and more widely distributed. Therefore, it is necessary to incorporate the equipment condition and risk assessment structure into the basis for decisions about the power distribution network dispatching system. For the above reasons, traditional SCADA, AM, FM, and GIS cannot fully satisfy the requirements for a modern DMS. The main problems are as follows.

1) A SCADA/AM/FM/GIS system can only monitor the macro operating characteristics of the equipment in the control system (such as oil temperature, current and circuit breaker open/close state). There is no function for on-line monitoring the insulation state of 35–220 kV electrical equipment (such as partial discharge, oil chromatography, or dielectric loss factor) in an urban electrical network. That is to say, the equipment insulation cannot be reflected in real time. So, the operation and maintenance personnel cannot obtain the relevant information in a timely manner.

2) SCADA/AM/FM/GIS systems do not have the function of fault diagnosis and cannot predict what kinds of fault will occur in the future, so it's hard to determine a proper period of condition-based maintenance or draw up proper dispatch pre-plans to properly address these problems [11, 12].

This chapter describes the technology of on-line monitoring of power supply equipment. After years of development, these technologies and systems came to be recognized and accepted by competent authorities on power transmission and transformation. At present, a cross-regional monitoring management system for transmission and transformation equipment is under construction by the SGCC. But most on-line insulation monitoring systems of distribution equipment under operation, which run independently, are still at the pilot study stage. As a result, they will not display the diagnosis in real time in a DMS system, let alone send out warnings to support DMS decision-making on the basis of a comprehensive analysis. Therefore, it is necessary to phase in a building condition monitoring system and a condition-based maintenance system in order to improve the operation and maintenance levels and strengthen the support for DMS decision-making.

6.1.1 Equipment Condition Monitoring Technology

6.1.1.1 Common Sensors

Sensors are important for the precision and reliability of the whole system, as a front unit for data acquisition in a condition monitoring system. Sensors are categorized into electric sensors and non-electric sensors. An electric quantity refers to an electric quantity in physics (e.g., voltage, resistance, capacity, or inductance); a non-electric parameter refers to parameters other than electric parameters (e.g., temperature, humidity, pressure displacement, or chemical components). Non-electric sensors convert non-electric physical quantities to quantities related to electricity; the measuring circuit measures, displays, and controls signals by processing and changing the electric signal output from the sensors. This section gives a brief introduction to the non-electricity sensors that are used in a condition monitoring system.

6.1.1.1.1 Humidity Sensor

Ambient humidity is often expressed as the density of vapor in the atmosphere – i.e., the content of vapor in $1\,m^3$ of air is often called the absolute humidity. Besides the absolute humidity, people's feeling of being dry or wet is related to the amount of saturated vapor in the atmosphere. The ratio of absolute humidity to pressure intensity of saturated vapor at a given temperature is called the relative humidity:

$$RH = D/D_s \times 100\% \tag{6.1}$$

where RH is the relative humidity, D is the absolute humidity in air (mmHg), and D_s is the intensity of saturated vapor pressure at the given temperature. It can be seen from equation (6.1) that when the intensity of vapor pressure in the atmosphere is equal to that of saturated vapor at the given temperature, the relative humidity is expressed as 100%.

There are many types of humidity sensor, such as macromolecule capacitive humidity sensor, lithium niobate crystal humidity sensor, quartz crystal humidity sensor, and thermistor-type humidity sensor.

6.1.1.1.2 *Temperature Sensor*

A temperature sensor is made using the characteristics of metal, semiconductor materials, and temperature, which include expansion, resistance, capacitance, magnetism, thermal electric potential, and so on. Temperature sensors are categorized into contact and non-contact sensors. The former have direct contact with the measured objects and the latter are installed to measure the temperature via the infrared radiation transmitted from the measured objects at a distance. The sensor can measure the local ambient temperature if located in the environment. At present, all types of temperature sensor are available, including platinum resistance, thermocouple, bimetallic strip, thermistor, transistor, and integrated circuit.

Owing to the advantages of good linearity and consistency, the integrated temperature sensor will integrate its sensor part, magnifier, signal processing circuit, and drive circuit into a chip, so the sensor can be widely applied in many fields for its small size and user-friendliness.

6.1.1.1.3 *Pressure Sensor*

Pressure sensors are most commonly used in industrial production and can be used to convert pressure into current or voltage and measure pressure and displacement. There are many types of pressure sensor, such as resistance strain gauge pressure sensor, semiconductor strain gauge pressure sensor, piezoresistive pressure sensor, inductance pressure sensor, capacitive pressure sensor, resonator pressure sensor, and capacitive acceleration sensor.

6.1.1.2 Distribution Transformer Condition Monitoring and Diagnosis Technology

6.1.1.2.1 *Transformer Failure*

There are many types of failure in transformers, and some are combined faults. According to our experience in maintenance and operation, the failures in transformers are basically categorized as internal failures (winding, iron core, internal assembly fittings, tap changers, lead, insulation oil) and external failures (oil tank, tap changer transmission mechanism, cooling unit, other accessories) by failure position. Failures are also categorized into sudden failures (insulation breakdown caused by over-voltage, winding deformation caused by external failures, short circuit, natural disaster) and failures caused by long-term accumulation (iron core failure, winding deformation, aging of insulation materials, suspended discharge assembly hardware fitting). Generally, internal faults can be determined from preventive tests and insulation oil chromatography analysis, while external faults can be located and determined by operation and maintenance personnel. Sudden failures cannot be predicted, since most failures are caused by external reasons and the faulty equipment is removed through differential, gas, over-current, and earth protection equipment only after the failure. Accumulative failures belong to internal failures. For failures other than sudden failures, it is possible to detect abnormalities and take countermeasures at an early stage to avoid serious failure or economic loss by the implementation of careful daily examination, preventive tests on a regular basis, and insulation oil chromatography analysis, especially the application of the latest technologies.

A wide range of failures may occur due to the design structure and manufacturing process of large power transformers, quality problems in installation, long-time operation, overload, operation, short-circuit impact, and lighting strike. In the event of failures in large transformers, widespread blackout will occur and may have a serious impact on industrial production and people's lives. So, once failures are found during the operation of a transformer, we should determine the causes of failure based on careful analysis, then locate positions and draw up a scheme of maintenance, so as to eliminate accidents from the beginning. There follow some common faults and an analysis of their causes [13].

1) *External failures and causes.* Failures that are commonly seen in transformers include oil leakage, explosion-proof vent and oil pointer breakage, fan, oil-submerged pump thermometer damage. These problems are visible from their appearance and can be handled by a minor overhaul. Table 6.1 lists the transformer failures and an analysis of their causes.
2) *Internal failures and causes.* Because of the complex inside structure, high voltage, and strong electric field, all varieties of failure may occur in a large transformer. For this reason, it is never an easy job to carry out cause analysis and make predictions. Therefore, careful research based on widespread experience is required. Generally, internal failures in transformers are categorized into:
   i) Circuit part – bushing, lead, winding, and tap changer.
   ii) Magnetic circuit – iron core, clamp, core-through screw, square iron, and shim plate.
   Table 6.2 shows failures on transformer parts and an analysis of their causes.

Table 6.1 Transformer failures and cause analysis

No.	Failures	Cause analysis
1	Abnormal rise in temperature during operation	Overload operation; too high ambient temperature; thermal devices not clean, or dirty; failure in cooling devices; radiator valve close; temperature indicator damage; internal failures in transformer.
2	Abnormal sound and vibration	Over-voltage; power grid frequency fluctuation; internal fastener is loose, or bad earth; suspended discharge; mechanical failure in cooling devices; metal part resonance; tap changer drive gear is defective; creep on porcelain surface.
3	Abnormal smell, discolor	Fastener loosen; too much heat on contact surface; overload; damp accessories.
4	Oil leakage	Aging gasket; poor welding; sand hole on metal part; bolt fastening.
5	Abnormal gas in other relay	Harmful free discharge causes insulation material aging; overheat on partial conducting parts; bad insulation in iron core; mechanical failures in oil-submerged pump.
6	Damage on porcelain surface	Discharge burn caused by over-voltage or pollution.
7	Explosion-proof equipment (or pressure release)	Assumed to be internal failure with relay protection device operation; respirator fails to breathe without relay protection device operation.

Table 6.2 Failures on transformer parts and analysis of their causes

No.	Type	Cause analysis
1	Failures in electric circuit	Failures that often occur to bushing of transformer, lead, winding, and tap changer are bad contact, discharge, or partial discharge, which can usually be detected by means of chromatographic analysis and preventive tests. In case of major failures, light gas or heavy gas protection will be triggered. Outage inspection is a must after heavy gas protection action. Trial operation is not allowed. Do not use again before the causes are found or suspended cores are inspected and dealt with.
2	Failure in HV lead-out bushing	HV bushing is usually oil-paper capacity-type bushing and for some main transformers, capacitive bushing is also used on LV sides. They often have failures of oil leakage, wetting, capacitive layer breakdown, and small bushing for grounding line breakage. The line breakage of small bushing for grounding is often caused by manufacture, transportation, or multiple times of connection/disconnection of ground wires.
		Failures can be detected by oil chromatogram analysis, bushing dielectric loss value, capacitive measurement, and bushing insulation resistance measurement test. If there is significant variation in measuring value or it exceeds the specified value, a correct judgment must be made based on comprehensive analysis. When the test shows that the results do not exceed the standard and the causes of defect are unknown, it is necessary to temporarily stop operation and shorten monitoring periods so as to better update and supervise faults.
		Treatment: When the bushing is seriously wet and the capacitive core breaks down, the bushing should be removed and returned to the factory for repair. Lift the small bushing down and remove it for soldering if the line breaks.
3	Failures in lead and winding	HV lead is sealing off, badly soldered, or open welded. Besides, there is poor connection at the joint between the lead and sleeve bolts, and it is oriented to short circuit in parallel. In a serious case, gas protection will be triggered. The fault is due to poor construction quality or winding deformation for external reasons.
		Fault detection: the fault can be detected through chromatographic analysis of insulation oil in the oil tank, or a DC resistance measurement test on the winding together with bushing.
		Treatment: first, inspect the pendant core or hoisting cover. When this is slightly overheated or there is an arc trace found on the breaker contact, it can be reused after repair. In more serious cases, replace the tap changer with a new one or repair the broken lead if it is burned.
4	Tap changer failure	Poor tap changer contact, short circuit or discharge to ground, tap changer lead is loose. Such failures make up the majority of faults, while the causes include lack of spring pressure, unreliable touch, insecure leads, oxidized breaker contact, tap changer not in place, caused by quality problems in installation, improper installation, operation, and maintenance.
		Fault detection: carry out oil chromatographic analysis on insulating oil in the oil tank and a DC resistance measurement test for winding together with bushing.
		Treatment: first, inspect the pendant core or hoisting cover. When this is slightly overheated or there is an arc trace found on the breaker contact, it can be reused after repair. In more serious cases, replace the tap changer with a new one or repair the broken lead if it is burned.

(Continued)

Table 6.2 (Continued)

No.	Type	Cause analysis
5	Magnetic circuit failure	Basically, we obtain dynamic information about the transformer continuously from all kinds of high-performance sensor located at the transformer. A smart software system and software rule program make it possible for condition-based monitoring devices to automatically monitor conditions. The condition-monitoring decision-making system is determined depending on variation trends in measurement parameters as time goes by, rather than absolute values of measured parameters. It works in a way that first collects, stores, and processes the data measured on site using high automation with a computer connected to the network, then makes a trend prediction. An integrated program for condition monitoring includes data acquisition and storage, state analysis and fault classification, locating the fault position based on experts' experience and proposals of the maintenance scheme.
6	Insulating oil quality degradation	The insulating oil electrical breakdown strength decreases, and dielectric loss increases, mainly because of being affected by damp or aging after a long time of operation.
		Detection: insulation oil assay, insulating oil dielectric test, withstand voltage test.
		Treatment: replace new oil and take oil protection measures. When there is an abnormality or abnormal content of characteristic gas in the insulation oil during relay action, external inspection, pressure-relief device action, and pretest. First, make a visual inspection, including oil injection, sound and location, state of relay actions, load and system operating conditions. At the same time, an electrical test should be conducted. Take samples from the oil tank, bushing, and gas relay, then judge the fault location and seriousness by a comprehensive analysis, which is vitally important for fault detection and recovery. If faults are not obvious enough to make a judgment, put the transformer into operation and keep track of it to update the information. After the locations and seriousness of faults are confirmed, if it cannot continue in operation because of the fault, the device should be terminated for maintenance. The methods or tests to detect faults and aging are insulation resistance and absorption ratio, winding DC resistance, transformation ratio, dielectric loss factor, capacitance, and insulting oil chromatographic analysis. So, to identify the seriousness of faults and locations in case of failure, gas-in-oil analysis is vital. When partial places are overheated, caused by internal faults, insulating materials, such as oil and insulting paper, would be heated, aged, and decompose to produce H_2, CO, CO_2, and hydrocarbon with lower molecular weight. These gases have large solubility, and large amounts of gas are dissolved in the insulting oil, but during early fault, a very small amount of gas is produced, and light gas relay protection is not triggered. Hence, it is unnoticeable. Since the insulating oil is circulating in the oil tank, gases are evenly dissolved. The method of gas-component analysis uses a mass spectrometer and gas chromatograph to analyze the gas components, gas content, and change year on year. The information is used comprehensively to identify fault types, locations, and seriousness. The method is able to determine minor failures and partial faults on the iron core that are difficult to find by electrical testing. Besides, the method can find potential faults at an early stage, but operators cannot forecast any abrupt incidence before insulation breakdown.

6.1.1.2.2 *Principle of Condition-Based Monitoring System*

Condition monitoring systems share basic principles, even if their monitoring items and goals are different. Basically, they obtain dynamic information about a transformer continuously from all kinds of high-performance sensor located at the transformer. A smart software system and software rule program make it possible for condition-based monitoring devices to automatically monitor conditions. The condition-monitoring decision-making system is determined depending on variation trends in measurement parameters as time goes by, rather than absolute values of measured parameters. It works in such a way that it first collects, stores, and processes the data measured on site with high automation via a computer connected to the network, and makes a trend prediction. An integrated program for condition monitoring includes data acquisition and storage, state analysis, fault classification, location of fault position based on experts' experience, and proposal of maintenance scheme [14].

Home and abroad transformer monitoring ranges are listed as follows.

- Hotspot monitoring with FOS (fiber-optical sensor).
- Measuring the total combustible gas in oil and analyzing the characteristic gas content of H_2, CH_4, C_2H_4, C_2H_2, C_2H_6, CO, and CO_2.
- On-line monitoring of partial discharge, including electric partial discharge, voice partial discharge, UHF partial discharge, static partial discharge.
- On-line monitoring of power factor and capacity of bushing.
- On-line monitoring functions of cooling devices (such as fan and oil pump transformation state).
- On-line monitoring of humidity, temperature, and acidity.
- On-line monitoring of load current.
- On-line monitoring of humidity and migration related to insulation paper.
- On-line monitoring of temperature at top and bottom winding.
- On-line monitoring of node characteristics and defects in electric system.
- On-line monitoring of clamping force of structural member.
- On-line monitoring of performances and defects of OLTC, including OLTC voice transmission, vibration in the OLTC tap-change operation, and drivability of OLTC motor.
- On-line monitoring of ground faults in the iron core and defects in winding.
- On-line monitoring of oil level in oil conservator, with oil leakage information provided by sensor.

    1) *Monitoring hot condition of transformer winding.* According to IEC regulations, the hotspot temperature of an oil-immersed transformer winding is limited to 118 °C. If the winding hotspot exceeds 140 °C during operation of the transformer, then CCO, CO_2, and water will be generated. But when the transformer winding is below 140 °C, there will be no characteristic gas. So, how to discover low temperature or overheating in a large transformer winding has become of concern for operation and manufacturing departments.

    Currently, it is the practice to monitor the condition of the winding by directly measuring hotspot temperature with a measurement sensor placed around the winding lead. The sensors used for measuring the winding temperature can be fiber-optical sensor or firefly fiber-optical thermometer. The key to applying technology to monitor conditions by measuring hotspot temperature lies in the layout and quantity. The new smart fuzzy sensor outputs numerical values directly after converting the measured values to those described in human language and studying, estimating, and deducing these numbers by reference to experts' theories and experience. This is how such sensors communicate and control. Both theoretical analysis and practical experiments suggest that such sensors work well in determining hotspot conditions.

Currently, in addition to the above parameters, a winding condition monitoring device can monitor deformation and life consumption per day/minute and display the calculated total life consumption of the transformer.

2) *Monitoring micro water in oil.* Water is one of the decomposition products of insulation oil below 500 °C. A substance called a "bridge" will be formed by water and impurities in the oil under the effect of an electric field, which breaks the strength of the insulation oil but reduces the resistivity and increases the leakage current. Therefore, it is significant for micro water-condition monitoring in oil to prevent the transformer being damaged by water and discovering oil-temperature faults below 500 °C. There are some errors in the traditional approach to monitoring the conditions of micro water in oil, the same as in other manual analysis approaches. However, all these errors can be eliminated by means of monitoring the conditions of micro water in oil. Such a device for monitoring the conditions of micro water in oil is an oil-immersed film-type polyimide capacitive humidity sensor, located in the transformer oil circuit.

Since polyimide is a new and stable moisture-sensitive material, its moisture adsorption capacity reaches 3.3% at 21 °C and RH of 100%, but both physical and chemical performances stay unchanged at −200 to 400 °C. So, the monitoring devices for micro water in oil make the best of polyimide materials and show their specific values at high oil temperature.

The working principles for monitoring the conditions of micro water in oil are as follows. When the oil flows through the sensor, the water contained in the oil will bring about changes in resistance and capacity at the polyimide film layer, causing the probe impedance to change. Through impedance changes caused by the dynamic balance relationship between film and water in the transformer oil, signals are received about humidity and temperature for the purpose of on-line monitoring the yield of micro water.

3) *Monitoring oil leakage.* Oil leakage has long been of concern for electric operation departments. To prevent the oil level from falling, due to oil leakage, and avoid insulation shutdown, the relevant authorities have developed an oil leakage-condition monitoring system. One basic function of the system is to keep track of oil leakage in the transformer. Such an automatic oil-leakage monitoring system presents variations in trend of the oil level in the oil tank and oil temperature data in the form of an electrical signal, converting and processing the monitored signals with a signal data converter. It should take the following steps to determine oil leakage: calculate the oil level according to the oil temperature and compare it with the actual oil level. A real-time automatic detection function brings convenience to the operation and maintenance personnel by saving manpower and resources, so an oil-leakage monitoring system is an important means to maintain the condition of the substation.

4) *Monitoring the condition of the iron core.* Multiple earthing faults in the iron core are commonly seen and make up a high proportion of transformer faults. If the earthing fault is not that serious, it is difficult to detect by means of chromatographic analysis. Measuring the earth current with a clip-on ammeter is still not that accurate.

5) *Monitoring the condition of containers in the oil.* In the event of overheating or discharge faults, the molecular structure will be destroyed and a large amount of hydrogen will be split from it. So, hydrogen contained in the oil can act as an indicator gas for primary faults. Besides hydrogen, there are certain amounts of other combustible gases, such as CH_4, C_2H_6, C_2H_4, CO, and CO_2. These combustible gases are mainly from the insulation oil and solid insulation made of organic insulating materials. Fission is likely to occur with molecular structures when the materials are subject to electricity, heat, oxygen, and water. For example, the oil would release H_2 and CH_4 above 500 °C, while the aging winding hotspot, insulation conductor wire, and insulation fiber component would also release CO and CO_2. A parts and materials factory burned by an electric arc would also release much H_2. Partial discharge would also produce H_2 and C_2H_2. There is a certain change in rules between combustible oxygen and time of operation from delivery to operation. Someone names the rule as a "fingerprint" in dissolved gas analysis. If there is an abnormal change in content of dissolved gas, this indicates that there are sources of characteristic gases usually produced by faults.

However, what should be noted is that it is still difficult to exactly determine the nature of faults – whether via monitoring the content of hydrogen or of combustible gas. Monitoring the gas dissolved in oil is supplementary and additional to gas chromatography (GC) technology and on-line monitoring of gas dissolved in oil is categorized into on-line chromatography, sensor, and infrared spectrum monitoring. Currently, we have a few methods to on-line detect dissolved gas, including air taking by film penetration, vacuum pumping, and carrier gas and air circulation.

The on-line monitoring of gas dissolved in oil is characterized by continuous observation of the dynamic development trend of gases. The system is an effective means to maintain a state, since it can discover gases that go beyond limits and gather information about faults to avoid any potential disaster. Characteristic gases differ. To determine whether there is an electromagnetism fault, we often use concentrations and concentration ratios of H_2, CH_4, C_2H_6, C_2H_4, C_2H_2, C_2H_4, CO, and CO_2. The ultrasonic signal received from the sensor is also used to determine whether there is a mechanical fault.

Monitoring of gas dissolved in oil could monitor a single gas and in whichever method, air is taken through various sensors and detectors for on-line monitoring. Gas–oil separation is key to monitoring gas dissolved in oil. This step makes a great difference to the results of monitoring, and uses an organic synthesis polymeric membrane for dialysis of faulty gas. So, the polymeric membrane in the detector should be of high performance. To begin with, it should be resistant to water, oil, and high temperature. Next, it should withstand mechanical damage. Finally, it should be highly sensitive to combustible gases.

Currently, Teflon™ film is the best among all gas-permeable membranes used in on-line monitoring devices, because it not only has high mechanical properties but is also oil and heat resistant. In addition, the Japanese Teflon B perfluorinated alkyl vinyl ether membrane has high performance as well, and H_2, CO, C_2H_2, C_2H_4, and C_2H_6 can permeate it.

It takes a couple or dozens of hours to get a balanced concentration of gas dissolved in oil via the permeating membrane. Meanwhile, the volume of permeating gas is determined depending on temperature. The higher the temperature, the faster the volume of permeating gas will increase.

The monitoring device for gas dissolved in oil is provided with a dissolved gas failure diagnosis system and a data analysis system. With a real-time concentration trend curve, if the data meet the conditions, then the monitoring device will automatically make a decision and give out audible and visual alarms. In the continuous process of monitoring, operators may control the concentration level for a gas alarm.

The monitoring device for gas dissolved in oil should be installed as close as possible to where the oil flows through, for convenience. Any delay in monitoring loses real-timeliness.

6) *Technology to monitor gas dissolved in oil.* This is a technology stemming from GC technology. Regular GC technology is a complicated and time-consuming test in the lab, and there are various errors caused by manual sampling and analysis. The development and application of oil CG technology remedies this defect in CG technology. China's transformer oil CG monitoring system adopts the CG principle and uses dynamic headspace (purge-trap) degassing technology and a microbridge detector of high sensitivity to detect seven components in oil. The system is integrated with CG, an expert diagnosis system, automatic control and communication technology, and on-line monitors the internal transformer with measurement and analysis of the gas dissolved in oil. Therefore, the system helps with the discovery and diagnosis of internal faults, and all are kept updated. The system is considered to remedy the defect of long maintenance period required in lab CG analysis and lay a solid technology foundation for ensuring safe and economical operation and condition-based maintenance. Figure 6.1 shows a schematic diagram of the working principle.

This system is characterized by high sensitivity, short analysis period, and consistent lab data, and is on-line chromatographic analysis in a real sense. After receiving the command, the on-line chromatography monitoring devices make a self-diagnosis after receiving the ON command from the main device (IED) and start up the environment, column box, degassing temperature-controlled systems, and finally start the oil circuit circulation system. The oil will flow into the instrument via a copper pipe from the transformer body and valve, and then return to the transformer body, flowing through another valve followed by a pump and oil return tank. After being stabilized as a whole, the oil in the degasser will be degassed with pump off. After repeated extraction by carried gas, the sample components are concentrated into a catcher and are expected to sweep into the chromatographic column rapidly with carried gas for the purpose of gas separation and detection. The residual oil in the degassing chamber is discharged into the oil tank. The detector will detect the concentration of each component and convert it into an analog signal proportional to the concentration, but limited to analog–digital conversion. These signals are sent to the master IED, which will transmit data via an Ethernet to a condition-based monitoring and analysis system for calculation and

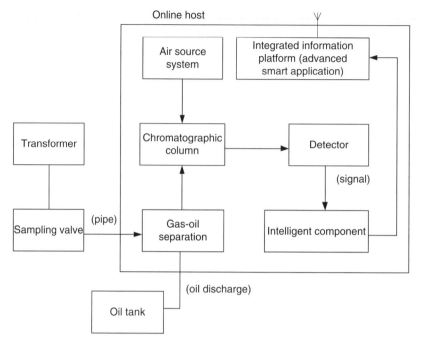

Figure 6.1 Schematic diagram of working principle of oil chromatography monitoring.

Table 6.3 Oil chromatography gas composition

Composition:	H_2	CO	CO_2	CH_4	C_2H_4	C_2H_6	C_2H_2

analysis, through which a trend chart of concentration variations will be automatically generated and provided to the smart diagnosis system for further diagnosis.

Description of condition monitoring devices: transformer oil will flow in circulation through an oil pump and transmit the separated gas to a dosing pipe in the flow valve. The gas samples in the reservoir pipe will be injected into a chromatographic column on a regular basis with the help of gases. With the electric signal limited to A/D conversion for all varieties of gas, input a single chip for processing and calculate the content of various characteristic gases. At last, the system will make an on-line diagnosis of transformer condition by way of characteristic gases, gas generation rate, and three-ratio methods.

Table 6.3 shows the gas compositions after the gases are separated from each other by a chromatographic column. The fault analysis will be based on the content of seven kinds of gas by computer.

The transformer oil chromatography condition-based monitoring expert system consists of a database, knowledge base, inference engine, knowledge acquisition, and man–machine interface. The main function of the database is to store and provide indicators that are reflective of changes in oil quality in real time and historic data. The indicator and information in the database further include analysis of defects in oil quality and results which could present details about the oil performance index to maintenance personnel. The knowledge base is to store and analyze experience and knowledge related to the transformer. The main function of the reference engine is to extract data from the database and make reference analysis of oil conditions in a logical form.

The condition-based analysis carries out and implements specific steps using characteristic gas analysis, three-ratio method, external test results, oil chemical and physical index, and insulation preventive test submodules. First of all, we calculate the absolute value and relative value of the transformer gas generation rate according to specific formulae. Second, we make a tendency analysis of oil decay combined with the water content in oil, acidity value, oil dielectric loss factor, and breakdown voltage. Third, we determine if there is a degradation or early failure in the transformer based on the analysis results.

Table 6.4 shows different gases for various faults. Tables 6.5 and 6.6 show the results of fault diagnosis by the three-ratio method.

Table 6.4 Gases from different types of fault

Type	Main gases	Minor gases
Oil overheat	CH_4, C_2H_4	H_2, C_2H_6
Oil and paper overheat	CH_4, C_2H_4, CO, CO_2	H_2, C_2H_6
Partial discharge in oil and paper insulation	H_2, CH_4, CO	C_2H_2, C_2H_6, CO_2
Spark discharge in oil	H_2, C_2H_2	
Arc in oil	H_2, C_2H_2	CH_4, C_2H_4, C_2H_6
Arc in oil and paper	H_2, C_2H_2, CO, CO_2	CH_4, C_2H_4, C_2H_6

Note: The wet gas or bubble in oil is possibly caused by an increase in H_2 content.

Table 6.5 Coding rule

	Coding for ratio range		
Gas ratio	C_2H_2/C_2H_4	CH_4/H_2	C_2H_4/C_2H_6
<0.1	0	1	0
0.1–1	1	0	0

	Coding for ratio range		
Gas ratio range	C_2H_2/C_2H_4	CH_4/H_2	C_2H_4/C_2H_6
1–3	1	2	1
≥3	2	2	2

Table 6.6 Fault type identification

Coding combination			Fault type identification	Cases (for reference)
C_2H_2/C_2H_4	CH_4/C_2H_6	C_2H_4/C_2H_6		
0	0	1	Low-temperature overheat (below 150 °C)	Insulated conductor overheats. Take notice of CO and CO_2 content, and CO_2/CO.
	2	0	Low-temperature overheat (150–300 °C)	Poor contact in tap changer; lead and split line bolts are loose or junction seal is bad; copper is overheated by eddy current; leakage flux in iron core; partial short circuit; poor insulation between layers; iron core is multi-point grounded.
	2	1	Intermediate-temperature overheat (300–700 °C)	
	0/1/2	2	High-temperature overheat (above 700 °C)	
	1	0	Partial discharge	High humidity and high content of gases leading to partial discharge of oil in low energy density.
1	0/1	0/1/2	Low-energy discharge	Lead discharge and flashover among parts with unfixed potential; connecting-tap lead and oil seam flashover; spark in oil among various potentials or floating on potential discharges.
	2	0/1/2	Low-energy discharge and overheat	
2	0/1	0/1/2	Arc discharge	Coil turn-to-turn; layer-to-layer short circuit; phase-to-phase flashover; tap-change lead-to-lead flashover; lead discharge to casing; coil fuse; tap-changer arcing; arc caused by circuit current; pilot wires discharge to grounding device.
	2	0/1/2	Arc discharge and overheat	

6.1.1.3 HV Breaker Condition-Based Monitor

6.1.1.3.1 Breaker Faults

According to statistics on breaker faults throughout the world, mechanical faults make up 70–80% of all faults, and electrical faults come second. In order to prevent failures from happening, minor repair or overhaul is carried out on breakers at intervals. So, it is necessary to maintain the breaker on a regular basis for safe operation, considering the various types of fault and frequent operations. The most serious fault for a breaker is failure in opening or closing – i.e., refusal operation or false tripping – and such a fault also does harm to the electric power system. A non-fault trip occurs in the breaker from time to time, mostly because the lowest action voltage for opening/closing the electromagnet or closing the contact terminal is set too low, and strong electromagnetic interference easily gives rise to a faulty trip signal. Electrical failure lies mainly in the contact electric loss, degradation in the arc-extinguishing medium, or substandard pressure or density, and most abnormalities are caused by a drop in performance of the vibration part of the machine (e.g., the voltage for opening or closing action is below standard; synchronization is not qualified; speed of opening or closing is not qualified; time of operation is too long). Faults are commonly seen in the operation part and the control circuit. In an arc-extinguishing chamber, changes in break time are often caused by contact burning and degradation in oil-blast medium performance.

Taking the SF6 breaker of a hydraulic actuator as example, Figure 6.2 shows almost all failures.

6.1.1.3.2   Basic Requirements for HV Breaker Condition-Based Monitoring

A breaker can exhibit a wide variety of faults. So, monitoring a single function does not satisfy the requirements of modern electric power systems for unattended operation in integrated automation. What's more, a single parameter is not a full reflection of healthy conditions. Neither is it good practice to achieve a simple superposition of all functions of single modules, because in so doing, the cost–performance ratio will be higher than desired. Besides, the measuring value is not made best use of, since one value may be used for multiple monitoring functions – e.g., coil current for opening and closing used to monitor not only the mechanical performance of a HV breaker, but also the integrity of a secondary circuit. The following items are supposed to be monitored with a view to monitoring a HV breaker all round, and realizing unattended operation in a real sense.

1) HV breaker electrical life – keep a record of the breaking current and arc time; assess the electrical durability of the HV breaker.
2) HV breaker mechanical life – the accumulative breaking times of the HV breaker under different classes of current.
3) HV breaker mechanical characteristics – current for breaker opening or closing coil; voltage of breaker opening or closing coil; breaker time-travel curve; vibration signal in operation of the breaker.
4) Integrity of secondary circuit – current and voltage of breaker opening and closing coil.
5) Insulation dielectric character on-line monitoring – breaker SF6 gas humidity and temperature; on-line monitoring of breaker SF6 pressure; water content in oil; oil gas analysis; tan δ, leakage current and temperature; vacuum degree.
6) Performance of actuator.

6.1.1.3.3   Analysis of HV Breaker Condition-Based Monitoring Principle

HV Breaker Mechanical Performance-Based Monitoring   The current waveform of the breaker closing or opening coil includes information about core strike, core stuck, coil conditions (short circuit or not), releasing, and fastening. Most importantly, we can learn the conditions at a specific moment in time. In addition, it is practicable to predicate the HV breaker when there is no-fault tripping or a refusal operation based on the analysis of the HV breaker operation mechanism through the coil current waveform.

The travel–time characteristic when the HV breaker is opening or closing is an important indicator of the mechanical characteristics, and the characteristics

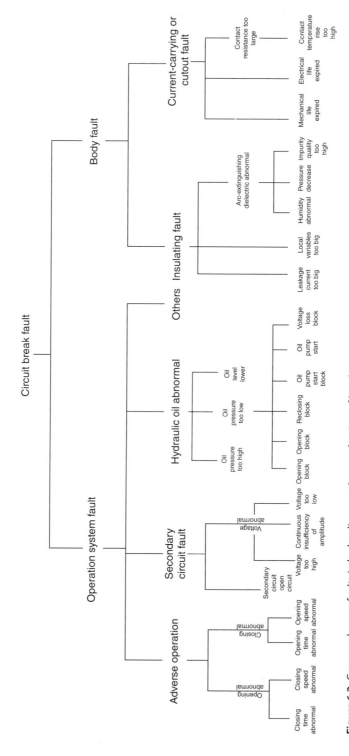

Figure 6.2 Commonly seen faults in hydraulic operating mechanism of breaker.

together with the time parameter will help people to calculate the speed of breaker closing and opening. Generally, a photoelectric encoder is most used for measurement in China.

Analysis of a vibration signal is a non-invasive monitoring means and the greatest advantage is that it does not involve any measurement with a sensor installed at the grounding part of the breaker, with no prejudice to normal operation of the breaker.

Taking the category of a spring, we could have a comprehensive understanding of the mechanical conditions of breaker multiple measurement values, by monitoring the current of the breaker opening or closing coil, DC operating power voltage, mechanical vibration signal, and HV breaker travel–time characteristics in a time domain.

Figure 6.3 shows the waveform of multiple monitoring values as the HV breaker opens. t_0 indicates the moment when the coil receives the command of opening, which is input through a switch value channel.

t_1 indicates the starting moment of opening the electromagnetic core, and $t_1 - t_0$ indicates the starting time of the core, which is closely related to mains voltage, coil resistance,

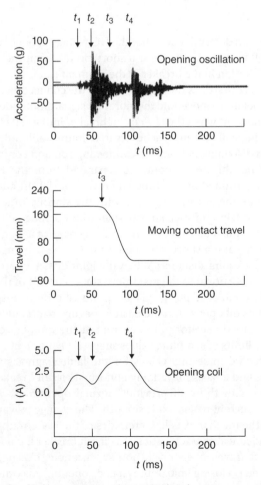

Figure 6.3 Waveform of multiple monitoring values as HV breaker opens.

inductance, and opening/closing electromagnetic irons idle stroke. The electromagnetic relation of the coil in the process of opening and closing is expressed as

$$U = Ri + L\frac{di}{dt} + i\frac{dL}{d\delta}\upsilon \tag{6.2}$$

where: $U/i/R$ is the voltage/current/resistance of the coil; L is the coil inductance; δ is the air gap in the electromagnet; v is the velocity of movement.

Within the starting time of cores, v is equal to 0 when the core current is stable and coil current i varies with the variation of index curve according to the formula

$$i = \frac{U}{R}\left(1 - e^{\frac{-t}{L_\delta/R}}\right) \tag{6.3}$$

The initial time is expressed as

$$t_s = \frac{L_\delta}{R}\ln\frac{1}{1 - I_d R/U} \tag{6.4}$$

As can be seen, we can determine whether the DC voltage of opening or closing the coil is 0, whether the coil is an open circuit or a short circuit, from the equation $t_s = t_1 - t_0$. There will be light vibration in the breaker at the moment t_1.

At t_2, when the iron core strikes the latched device, the connecting rod disconnects, the HV breaker contact begins to move, and the vibration signal amplitude reaches maximum.

t_3 represents the moment when the HV circuit breaker initiates its transmission mechanism and the opening coil current reaches its maximum stable value. At this time, the tripping mechanism disconnects from the connecting rod and contact is set in motion. After a response time, the main contact is separated from the HV circuit breaker. Traditionally, t_s is determined as the time when the maximum current i_{2max} appears, which is deemed to be the arc starting time with the starting time ignored. $t_1 = t_3 - t_2$ indicates the status of trip and mechanism when the opening coil is at a stable voltage – i.e., at the maximum stable voltage. If the iron core has a longer idle travel, it will get more kinetic energy and a shorter tripping time t_1. The tripping time t_1 is inversely proportional to the lock joint distance. When the joint distances are fixed, the tripping time t_1 is inversely proportional to the maximum stable current of the opening coil.

When the moving contact is put in place, t_4, the vibration caused by the buffer is stopped, the breaking coil opens, and the current begins to attenuate.

Seen from the recorded waveform, the current at the moment of each event has a limit. In this way, the limits of sampling values suggest the moment of each event, which can be taken as a record in setting values for the fault diagnosis software as well as characteristic values and a basis for judging abnormal circuit breakers. Comparing the measured waveform with those of previous normal operations, we can restore the waveform of the coil current to see if it is smooth. The change in waveform current can determine whether the mechanical characteristics of the HV circuit breaker are stable.

Generally, the mechanical characteristics of a circuit breaker include the opening time, closing time, contact travel, over-travel, non-synchronous closing, non-synchronous opening, closing speed, closing maximum speed, opening maximum speed, average closing speed, and average opening speed.

The mechanical characteristic parameters are defined as follows.

1) Closing time – the time interval from the moment the closing circuit is energized to the moment when all polarities of main arc contacts are contacted.
2) Opening time – the time interval between the moment when the current begins to flow through the opening circuit and the moment when all polarities of main arc contacts are separated.
3) Contact travel – the total movement between position differences of moving contacts.
4) Over-travel – the displacement when all contacts start touching until the closing position.
5) Closing non-synchronous – the time difference between the first polarity and the last polarity.
6) Closing non-synchronous – the time difference between the first polarity and the last polarity.
7) Closing speed – the average speed within 10 ms after the closing auxiliary contact is energized.
8) Opening speed – the average speed within 10 ms after the opening auxiliary contact is energized.
9) Closing maximum speed – the maximum speed in the process of closing.
10) Opening maximum speed – the maximum speed in the process of closing.
11) Average closing speed – the average speed for 80% of travel in the operation of closing.
12) Average opening speed – the average speed for 80% of travel in the operation of opening.

To get the above parameters, we should have moments and contact displacements within the period of time to calculate the speeds in an indirect way. To make the mechanical characteristic parameters more available, a closing and opening circuit break is considered as a sequential time chain linked by an action process, and each action is a subsequent time chain. To monitor the process of the circuit breaker, we must find a way to capture the relative position of the moment of operation of parts of the circuit breaker in the sequential time chain, and then the time parameters will be clear. After this, all kinds of characteristic parameters can be calculated by microprocessor. Since a split-phase operation of the circuit breaker involves no problems of opening/closing synchronization, it is very simple to analyze. To illustrate this, process time sequence diagrams for the three-phase synchronous operation of circuit breaker opening and closing are shown in Figures 6.4 and 6.5, respectively, and then the characteristic parameters can be expressed by parameter points in the figures.

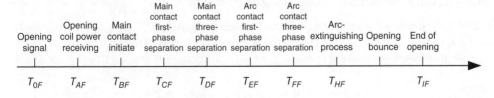

Figure 6.4 Sequence diagram for opening circuit breaker.

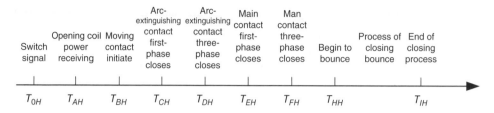

Figure 6.5 Sequence diagram for closing circuit breaker.

Assume the closing and opening speeds to be $V_H(t)$ and $V_F(t)$, respectively.

Opening time:	$T_F = T_{FF} - T_{AF}$
Closing time:	$T_H = T_{FH} - T_{AH}$
Time for opening out of synchronization:	$T_{FBTQ} = T_{DF} - T_{CF}$
Time for closing out of synchronization:	$T_{HBTQ} = T_{DH} - T_{CH}$
Contact total travel:	$D = \int_{T_{BH}}^{T_{HH}} V_H(t)\,dt$
Excessive distance:	$D_b = \int_{T_{EH}}^{T_{HH}} V_H(t)\,dt$

HV Circuit Breaker Electrical Life   Opening or closing a circuit breaker does damage to the circuit breaker (and more to the contact) to some extent, so the degree of wear of the contact becomes one measurement gauge of circuit breaker performance. Generally, the degree of wear of the contact in cut-off is used as one index to measure the electrical life of the circuit breaker. There are shortcomings in the traditional method based on HV circuit breaker life curve and breaking current: first, it ignores the differences in three-phase breaking current, especially that between faulty and non-faulty phases; second, it doesn't resolve the problem of three-phase asynchronization for the post-open phase with longer arc time and larger electrical wear; third, the determination of the arc time, especially the starting time of the arc, directly influences the electrical wear of the contact.

Electrical devices are capital-intensive equipment, so controlling the life loss to prolong the life and make the best of the remaining life is important. Generally speaking, the life of the equipment and system is determined depending on the parts with the shortest life. Since a HV circuit breaker is a switching current device, its electrical life is usually shorter than the average life of a HV circuit breaker. Hence, it is also important to carry out on-line condition-based monitoring and diagnosis of circuit breaker electrical life.

Currently, many researchers focus on electrical life. There follows a brief introduction to some common methods of calculating electrical life.

1) *Accumulative breaking current.* Since the degree of arc burning has much to do with the current, it is possible to reflect the burning conditions of the contact by means of an accumulative breaking current. We obtain the results according to the calculation formula

$$Q = \sum_{i=1}^{N} I_i \tag{6.5}$$

where Q is the total electrical wear value, I_i is the breaking current, and i is the time of breaking. With an accumulative current before maintenance as initial value provided by the circuit breaker manufacturer, we can calculate the remaining life margin. In respect of the average breaking time in a large number, like 20 and 30, there is no great difference in average breaking time. Therefore, the accumulative breaking current is widely used at the first stage of monitoring electrical life. Still, there are some shortcomings. In fact, the degree of burning may vary even when a circuit breaker breaks twice under a current of the same magnitude with the same external conditions. When there is a large difference in breaking current, the burning principles of contact are different, so it is far less accurate for the accumulative breaking current to determine the actual burning degree of the contact. Since many practical issues are ignored in the algorithm, this method lacks precision.

2) *Accumulative arc energy.* Since the degree of arc burning has much to do with the arc energy, an accumulated arc energy representation is more accurate than merely an accumulated breaking current value. The values are calculated according to the formula

$$Q = \sum_{i=1}^{N} I_i^2 t_i \tag{6.6}$$

where Q is the total electrical wear, I_i is the breaking current value, N is the number of breaking times of the circuit breaker, and t_i is the breaking time. The algorithm is so simple that all we need are the measured values of arc time and breaking current. However, in practice, it is not easy to measure the arc time, so we often replace it with the breaking time.

3) *Method of relative electric wear.* Whatever type of circuit breaker we have, we can learn about its allowable breaking time at a rated breaking current from the product introduction. Let us assume that the abrasion loss of contact for a single breaking under a rated current is M, and the allowable number of breaking times is N, so the total abrasion loss $Q = NM$. When the breaking current is as small as 3% of the rated breaking current, the abrasion loss amounts to that for breaking with full capacity, which is so little that a breaking current smaller than 3% will be considered a 3% breaking current. First, we define the allowable contact wear for a brand new circuit as 100% (i.e., relative life equal to 1), so the relative abrasion loss for each break at a rated breaking current is $1/N$. Combined with the $N-1$ curve of the specific circuit, we get the allowable breaking time N at any current and then the corresponding electrical abrasion loss from the equation $Q_m = 1/N$. In so doing, we get the relative abrasion loss at any current as well as the relative life of the circuit breaker:

$$L = L_1 - \sum Q_m \tag{6.7}$$

where L_1 represents the initial value of the circuit breaker life, a percentage smaller than 1 which is determined by the operation history. So, for a circuit breaker that is first put into operation or goes through overhaul, L_1 is equal to 1.

4) *Accumulation method of weighting breaking current.* The most widely used method to monitor electrical life is the accumulation method of weighting the breaking current of an electric curve (N_1–N_c). Generally speaking, the electric life curve is illustrated by

$$N_1 I_c^\alpha = Q_N \tag{6.8}$$

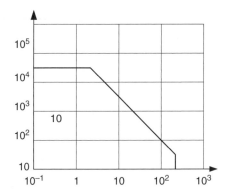

Figure 6.6 Curve chart of electric life for some circuit breaker.

where α represents a weighting index of the breaking current, varying from 1 to 2; l_c is a valid value of the breaking current; N_1 represents the allowable breaking time when the valid value of the breaking time is equal to I_c; and Q_N is allowable point wear.

Taking logs on both sides of the equation and consolidating them, we have

$$\lg N_1 = -\alpha I_c + \lg Q_N \tag{6.9}$$

As we can see from equation (6.9), $\lg N_1$ is a linear function of l_c. The curve chart of electric life for a given circuit breaker is illustrated in Figure 6.6. Taking two different coordinates from the curve, we obtain the straight slope $-\alpha$ and then Q_m. If I_c^α is used to express the electric abrasion loss when the valid value of the breaking current is I_c, then according to equation (6.10) we have the accumulated electric loss Q_1 of the circuit breaker in operation:

$$Q_1 = \sum_{i=1}^{n} I_{ci}^\alpha \tag{6.10}$$

Here, n represents the number of times a circuit break occurred.

The electric life of the contact can be taken from comparison of Q_n and Q_1. I_{ci} in equation (6.10) represents the valid value of a breaking point, which can be taken from

$$I_{ci} = \sqrt{\frac{1}{T_o} \int_0^{T_0} \left(i_{ci}\right)^2 dt} \tag{6.11}$$

where i_{ci} represents the ith breaking time and T_0 represents the breaking time.

6.1.1.3.4 HV Circuit Breaker Insulation Dielectric Performance Monitoring

Monitoring SF6 Temperature and Pressure   For an SF6 circuit breaker, SF6 is not only an arc-distinguishing medium but also an insulation medium and mainly measures water content and pressure. Water content will have a great influence on the components and content of decomposition products, since it produces corrosive acid gases like HF and SO_2, which cause contact corrosion deterioration. Humidity must be measured as accurately as possible, since the testing temperature of 20 °C, as well as the humidity, are specified in the relevant standards. The humidity of SF6 changes with the ambient temperature, and the temperature variation lags behind the ambient temperature. Therefore, the humidity value measured below 20 °C, in a strict sense, does not stand for the gas humidity at high temperature. The measurement of humidity cannot be made without the measured value of the temperature. Besides, both the humidity and temperature must be taken into consideration. We can determine if there is slow leakage of SF6 by monitoring the SF6 pressure.

Monitoring Insulation Oil in Oil Circuit Breaker In the closing of the circuit breaker, the oil disintegrates into H_2, C_2H_2, CH_4, C_2H_4, and C_2H_6, and these gases are dissolved in the oil and affect the arc-extinguishing and insulation properties if their content is too high. For a bulk-oil circuit breaker, there is an extra insulation test for the insulation oil of the arc-extinguisher chamber. With the rise in oil temperature, tan δ and I would change, so the changes in temperature, tan δ, and I are a full reflection of the insulation of oil. In addition, tan δ of the electrical insulation is closely related to temperature, which varies from material to material and structure to structure. Generally, tan δ increases with rising temperature. Since the temperatures are variable in the field, so the temperature should be included in the measurement of tan δ, so that all measured values of tan δ can be converted to values at the same temperature for better diagnosis.

Monitoring the Degrees of Vacuum for a Vacuum Circuit Breaker Once the degree of vacuum in a vacuum arc-extinguishing chamber of a vacuum circuit breaker decreases for some reason, the internal flashing voltage will be as illustrated in Figure 6.7. The degree of vacuum is monitored using this phenomenon. Among the available methods, including addition voltage, discharge current detection, discharge interference monitoring, intermediate potential change monitoring, and direct monitoring, the last two are most applicable. The method of middle potential change monitoring works in such a way that when the protection screen potential changes as the degree of vacuum reduces, a capacitance is connected to the protection screen by which the discharge voltage is monitored. We can determine the change in degree of vacuum according to the variation trend of the voltage, or measure it directly by placing a detector in the arc-distinguishing chamber.

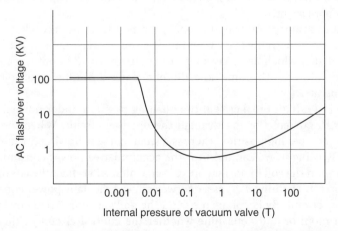

Figure 6.7 Curve relation between vacuum arc-distinguishing chamber and AC flashing voltage.

Problems   All these methods have the same problem, which is that the integrity of the closing circuit loses monitoring when the circuit breaker is closed, as well as the integrity of the tripping circuit once the circuit breaker trips, so not all conditions are under supervision. The idea has been put forward to monitor the condition of the opening or closing circuit and operation performance with the use of secondary operation work. This so-called secondary circuit operation work measures the power consumption by closing (opening) the coil in the operation of the HV circuit breaker. We obtain the secondary circuit closing operation work E_c according to the formula

$$E_c = \int_0^{t_c} u_c i_c dt \qquad (6.12)$$

where u_c is the closing coil voltage, i_c is the closing coil current, and t_c is the conduction time for closing the coil, from the receipt of power to the loss of power.

The secondary circuit opening operation E_o is

$$E_o = \int_0^{t_o} u_o i_o dt \qquad (6.13)$$

where u_o is the opening coil voltage, i_o is the opening coil current, and t_o is the induction time for closing, from the receipt of power to the loss of power.

The secondary operation work is affected by the following factors: movement of iron core; condition of transmission mechanism; switching time of HV circuit breaker auxiliary contact; conditions of opening or closing coil. That is to say, the struck iron core, increases in length of travel, in switching time of HV circuit breaker, short circuit in coil turn-to-turn, and reduction of grounding insulation in operation circuit would lead to increased operation power of the secondary circuit. Under normal conditions, the operation power for the opening or closing secondary circuit stays unchanged. The working power for the secondary circuit will not drop while there is a possibility of a drop in operating power of the opening secondary circuit. When E_o has a tendency to reduce, this suggests that there is a high possibility of a no-fault trip for the HV circuit breaker. Determination via trend is a characteristic of monitoring the operation power of the secondary circuit.

The greatest advantage of using the operation power of the secondary circuit as an indicator of conditions in the HV circuit breaker operation circuit is that conditions are recorded in digits, which are convenient to compare. In addition, since current and voltage are monitored at the same time, this eliminates the influence of voltage fluctuation in the operation.

When the HV circuit breaker is at the opening position under normal conditions, there is a small current flowing through the opening circuit. When the HV circuit breaker is at the opening position, there is a small current flowing through the opening circuit. By using it, we can monitor the accumulative power consumption of the circuit breaker on the coil in normal operation, which will reflect the conditions of the secondary circuit. Defining T_c/t_o as a day, we can find the daily power consumption of the secondary circuit. By keeping an eye on the daily accumulative power consumption, we can monitor and determine whether the control circuit of the HV circuit breaker is intact in real time. The system gives out warnings when the circuit breaks a line, or in case of power supply disruption, short circuit of opening/closing coil, or open circuit.

When the HV circuit breaker is closed and the dynamic breaking auxiliary contact on the HV circuit breaker is energized, DC operation power adds to the relay coil through the closing position with a large resistance to the opening coil; in the process, some power is consumed. In normal conditions, there is little change in coil operation power. When a turn-to-turn short circuit or reduction in grounding insulation occurs on the opening coil, the power consumption increases greatly. When there is poor contact between the HV circuit breaker auxiliary contacts, the pressure drop increases, the voltage level decreases, and the power consumption increases. The power consumption will reflect the mains voltage and play an important role in monitoring the mains voltage. The same applies with the closing circuit.

Classification of Circuit-Break Monitoring   To monitor the condition comprehensively, a lot of values need to be measured. A circuit breaker with various actuators has comment-monitoring variables as well as own measuring variables, as shown in Tables 6.7 and 6.8.

In the monitoring of pressure and degrees of vacuum, we could use the pressure sensor to input signals in the range 4–20 mA; humidity should be measured with its own generation and transmission units, and temperature is measured with a self-regulating sensor. Gas sensors are used to measure the content of gases in oil and input current signals ranging from 4 to 20 mA.

6.1.1.4   Lighting Arrester Condition Monitoring

6.1.1.4.1   *Lighting Arrester Faults*

In the 1960s, Japan took the lead in developing a zinc oxide nonlinear resistance chip. With the advantages of low residual voltage, non-flow current, low time delay, and large capacity of through current, the metallic oxide arrester (MOA) composed of ZnO nonlinear resistance chip replaced SiC arrest and was widely applied and rapidly developed

Table 6.7 Public monitoring variant

Public monitoring variant	Sensor	Public monitoring variant	Sensor
Current and voltage for opening or closing coil	Hall element	Speed of closing or opening	Indirect acquisition
Opening or closing time	Indirect acquisition	Speed of closing or opening	Indirect acquisition
Asynchronous time of opening/closing	Indirect acquisition	Breaking current	Current-transformation converter
Speed of opening or closing	Indirect acquisition	Busbar voltage	Potential converter
Travel-time characteristics	Displacement sensor	Insulation-pole leakage current	Self-made transformer
Vibration signal	Accelerator sensor		

Table 6.8 Monitoring variant of characteristics

Characteristics/type	Oil	SF6	Vacuum	Operating mechanism		
				Pneumatic	Hydraulic	Spring
Oil level	✓					
Oil gas	✓					
Dielectric loss	✓					
SF6 pressure/density		✓				
SF6 humidity/temperature		✓				
Vacuum degree			✓			
Rated operation pressure				✓		
Opening/closing lock pressure				✓		
Safe valve working pressure				✓		
Compressor running time				✓		
Rated working oil level					✓	
Pre-charge nitrogen pressure					✓	
Opening/closing lock oil pressure					✓	
Safe valve action oil pressure					✓	
Oil pump running time					✓	
Spring acting movement						✓
Spring storage time						✓

in power systems. Currently, the MOA has become the most popular and developed over-voltage protective device in power systems.

A 10-kV distribution network is a link with the most complicated electric power structure and environment. A vast number of distribution transformers and cable lines are equipped with arrest for lighting protection. Therefore, arrester faults are common in 10-kV distribution networks. These faults mainly lead to hazards as follows:

1) The arrester is not broken down completely, the leakage will do harm in line loss, which is not good for the economic operation of the power grid either.
2) The arrester is broken down, leading to a one-point grounding fault. Since the arrester fault is a hidden fault, it requires a lot of manpower and materials to locate the fault. If two phases on the arrester are broken down, respectively, two-point grounding faults will trigger switch protection action and lead to power interruption for consumers.
3) An arrester explosion will affect other devices in the surrounding area and make the accident more serious.

So, in the security assessment, we have attached equal importance to monitoring the arrester in the distribution system and monitoring the arrester in the power plant or substation. Meanwhile, since distribution network arrest costs are low and have minor influence, the monitoring of the arrester has not been taken seriously. In China's power system, we usually monitor the arrester on a two-year cycle in a preventive test method by taking the arrester apart for testing. Because of the large number of arresters in a distribution network, it costs a lot in terms of manpower and materials, and power outage occurs in some cases. At the same time, there is a possibility of damaging the arrester during the process of repeated preventive tests. So, more than one power utility has modified the test cycle to three to five years, and even cancelled the preventive test until replacing parts when the MOA is aging or wet. All these practices increase the risk of security hazards. So, it is imperative to monitor the conditions of the arresters in a distribution network.

An arrester in a distribution network usually adopts housing made of composite insulating materials, so it is difficult to find faults once a short circuit is arrested. Therefore, it is not possible to keep up maintenance and normal operation of the distribution network. This is an ideal means to monitor conditions under operation. However, there is little research on the monitoring conditions of distribution networks, which can be explained by the facts listed below.

1) *Limits in monitoring methods and costs.* The arrester for a distribution network is only available to monitor leakage, what's more, the cost in system voltage is too high. For these reasons, arresters are not popular or widely applied in distribution networks.

2) *Hard to extract leakage current.* The leakage current in the normal operation of an arrester in a distribution system is negligible, at only a couple of milliamperes, which an ordinary sensor will fail to detect. But in recent years, the growth in current sensors has made it possible to detect current at the milliampere level.

With the rapid growth in industrial power, there are higher requirements for reliability of generation and transmission devices. In recent years, the MOA has been increasingly widely applied in the field of over-voltage protection. Whether a domestic or foreign MOA, there are cases where the MOA exploded or had defects. Examples of accidents are: four 330-kV MOAs occurring at Longyangxia Station in 1989; a 500-kV MOA in Japan in 1991, 1993, and 1994 at Gezhouba Station; a 220-kV MOA found wet at Wu River Hydroplant in 1994. These accidents have increased attention on the operation and manufacture sectors. According to statistics, since the application of zinc dioxide arresters in power systems in the 1980s, the accident rate of the Chinese zinc dioxide arrester at the 110-kV level or above has dropped to 0.68% and the exit rate to 1.3%.

6.1.1.4.2 Classification of Arrester Faults

Arrester faults are divided into four classes.

1) *Faults caused by interior damping in arrester.* According to historic accident records, the faults caused by interior damping account for 60% of all arrester faults. What's more, seen from the wreckage of damage affected by damping, the characteristics of a damped arrester are:
   - flashover trace on the external side of the valve block and the inside of the insulator;
   - no trace of discharge on the aluminum surface of the valve block;
   - the leakage of current increased significantly before the accident;
   - the pre-accident leakage current doubled;
   - the insulation resistance decreased greatly.

The sources of humidity are mainly:

- the humidity exceeds the value specified by the manufacturer of the arrester;
- the valve block and inside components are not dried completely and moisture is left over;
- the seal ring is misplaced in the assembly process or there are sundries between the seal ring and the sealing surface of the insulator.

2) *Accidents caused by deterioration of arrester as it ages.* Some manufacturers seek to improve the protective performance and advanced index blindly, ignoring the reliabilities of the arrester. Too low a residual voltage and the rated voltage may cause the electric load rate and load to increase. For a HV arrester, due to its height and size, it is vulnerable to the environment. Also, because of the uneven distribution of potential, the chargeability of some parts has reached the tolerable limits of a metallic dioxide resistance chip, which in turn accelerates the aging of the local resistance chip and leads to a change in volt characteristics of the entire arrester, a deviation from the thermal stability working point, and an increase in temperature of the chip. Because of the negative temperature coefficient of the resistance chip under the working voltage, U_{ImA} will reduce more as the temperature rises. At this time, U_{ImA} is close to the peak voltage in normal continuous operation. Once the grid voltage exceeds the power frequency withstanding voltage, the arrester will be damaged. An example is the 18 arresters at 110, 220 kV that have been in service for five years and were involved in a succession of accidents in 1990. Five MOAs were damaged, while for the 13 undamaged MOAs, their average leakage current increased by 92% under the operating voltage and the valve block was seriously aged. In a similar case, a 500-kV arrester from Sweden was out of service after two years since it was found that the potential was highly distributed at the upper unit with aging, attributed to the design elements of high chargeability and uneven distribution of local units.

3) *Arrester damage caused by environment and pollution.* Years of field monitoring and artificial pollution tests by power utilities have proved that pollution and environment are major contributing factors in arrester aging. Influenced by high temperature and pollution, the potential distribution is very uneven. It is indicative of a high load rate and accelerated aging at places near the flange with high temperature and large current flowing through. Years of survey show that accident-prone areas are hot and humid areas in South China and polluted areas in summer. For instance, among nine sets of 330-kV MOAs that have been in service for more than two years in Longyangxia power plant, four were damaged in succession during the period of drainage via the bottom hole. Later, the on-site analysis found that spray produced a layer of random changeable water film with small water resistance, and it was the water film that led to uneven distribution of arrester potential and finally damaged the devices. Of course, we do not rule out the possibility that the arresters were not well enclosed.

4) *Arrester damage caused by abnormal operation conditions and other factors.* Anything may happen to an arrester, but some events are beyond the tolerance range – such as a strong earthquake, direct lightning flash, or resonance. It is almost impossible for arresters to avoid damage under these circumstances. It is also likely that improper operation may cause a neutral grounding system to change into a neutral insulation system. Arrest may be damaged after working for a long time at high voltage. For instance, two arresters in Beijing power plant were once damaged by improper

operation. In addition, arresters may be damaged due to resonance caused by failure of a circuit breaker and phase-deficient operation. In another case, cracks appeared on a valve block because of a loose connection in transit, and later a resistor disc broke after running for a period of time. Thus, the arrester was out of order.

Under the operating voltage, the zinc-oxide valve plate core column, insulating rod, and porcelain bushing that compose the MOA all have a continuous flow-through current, called the MOA's full leakage current. With a dry and clean surface of the porcelain bushing, the leakage current flowing through the porcelain bushing and the insulating rod is far less than that from the MOA valve column. So, under normal conditions, the full current is mainly the leakage current passing through the MOA core column. The unevenly distributed pollutants, if any, on the surface of the MOA may lead to uneven potential distribution, so the leakage current on the porcelain bushing varies depending on pollution digress. For multiple MOAs, the arrester core body with a small leakage current is supposed to carry a heavy current. At the time, the power load will increase, which accelerates aging and increases heat stress. Thermal breakaway of the valve plate tends to cause the MOA to break down. When the MOA is poorly sealed and gets affected by damp, the leakage current will increase and the insulation resistance reduces. In that case, accidents are caused by flashover along the damped inner wall or valve shaft and subsequent earthing short circuit. Owing to the quality problems of the zinc-oxide valve itself, the valve tends to be thermally aged, and increased power consumption at normal voltage causes it to become hot. Thus, accidents happen because of valve thermal breakdown. The electric field around the zinc-oxide valve, together with the relative air humidity, would partially discharge the electricity, leading to valve aging. This is because a small amount of water and partially discharged electricity enables air decomposition to interact on zinc-oxide particles, causing the grain-boundary potential barrier to reduce.

MOA events result from many factors, in some cases as the combined action of the above. From the aspect of monitoring, an MOA's running characteristics will change as follows.

1) When the surface of the MOA porcelain is contaminated, on humid or rainy days, an abrupt change will occur to the full current and the resistive leakage current will be greatly increased.
2) When the MOA internal valves are affected by damp, under normal operation voltage, the fundamental harmonic component of the MOA resistive current increases greatly, whereas the high-order harmonics component increases slightly.
3) When the valve of the MOA ages, under normal operating voltage, the fundamental harmonic component of the MOA resistive current increases slightly, whereas the high-order harmonic component increases greatly, caused by the worsening nonlinear characteristics of the valve block.
4) Generally speaking, when the local discharge is more than 250 pC, there must be something wrong in the internal arrester. The order of magnitude of the local discharge impulse is microseconds, so it is necessary to test a high-frequency component. Many MOA events have proved that simply measuring the full and resistive current is not reliable. In these events, no exceptions were found in the measured parameters but accidents occurred later. Therefore, it is essential to explore the monitoring parameters and research a more reliable approach.

6.1.1.4.3 *Selection of Leakage Current Sensor*

The leakage current sensor is used to measure the leakage current. Since the leakage current has something to do with dielectric loss, the dielectric loss can be found through accurate measurement and analysis of the leakage current.

Since the current varies in a wide range (from several to hundreds of microamperes), sensors are required to have a wide enough dynamic range. What's more, the wide frequency spectrum included in the insulator partial discharge pulse signal determines that sensors should be provided with a wide frequency band (from a few to dozens of hertz), good characteristics of transient response, and linearity. In addition, for the sake of installation, it had best use current sensors that are sensitive to opening and closing.

Such a current sensor is usually one turn of the coil at the primary side. It would be more rewarding, where appropriate, for multiple turns to be used. Harness the round or opening square magnetic cores onto the grounding wire of the equipment, as shown in Figure 6.8. The selection of core materials is made depending on frequency of use. Ferrite can be used to measure high-frequency/pulse current. Manganese zinc ferrite's maximum useful frequency is 3 MHz (PW5), and relative magnetic conductivity 105. Recent years have witnessed the rapid development of a crystallite magnetic core. Its magnetic conductivity is above 104, so it is highly sensitive. It is easy to form by processing, and has a frequency varying from 40 Hz to 500 kHz, covering almost all frequencies in between.

The relationship between current signal $i_1(t)$ and inductive voltage on the ends of the secondary coil [i.e., the input signal $e(t)$] is expressed as

$$e(t) = M \frac{di_1(t)}{dt} \tag{6.14}$$

$$M = \mu \frac{NS}{l} \tag{6.15}$$

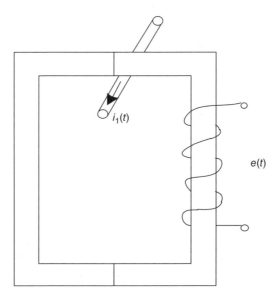

Figure 6.8 Current circuit-breaker structural schematic diagram.

where M is the inductive coefficient, N is the secondary coil number of turns, S is the section area of the magnetic core, and l is the length of the magnetic circuit. As we can see from equation (6.14), the output signal $e(t)$ is positive with regard to rate of change $i_1(t)$.

The current sensor is structured like a Rogowski coil used to measure heavy current impact, so it is also called a Rogowski coil. What differentiates it from a true Rogowski coil is the fact that a Rogowski coil is used to measure tens to thousands of kiloamperes, which is not that sensitive, and it uses an empty coil, instead of a magnetic core. Meanwhile, current sensors are used to sense small currents, varying from several milliamperes to microamperes. Both of them rely on the same principle.

Sensors are integrated in two ways, designed for wideband and narrowband sensors, respectively. Wideband sensors are described below.

A parallel wideband (also self-integral) current sensor connects a self-integral resistance R at the ends of the coil, as shown in Figure 6.9.

We have the equation

$$e(t) = L\frac{di_2(t)}{dt} + (R_L + R)i_2(t) \tag{6.16}$$

$$L = \mu\frac{N^2 S}{l}$$

where L is the coil inductance and R_L is the resistance of the coil. When the following conditions are satisfied:

$$L\frac{di_2(t)}{dt} >> (R_L + R)i_2(t) \tag{6.17}$$

we have

$$e(t) = L\frac{di_2(t)}{dt} \tag{6.18}$$

According to equations (6.16) and (6.18), we have

$$i_2(t) = \frac{1}{N}i_1(t)$$

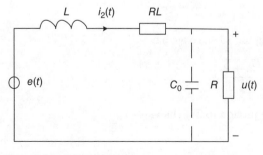

Figure 6.9 Equivalent diagram for wideband sensor.

so

$$u(t) = Ri_2(t) = \left(\frac{R}{N}\right)i_1(t) = Ki_1(t) \tag{6.19}$$

where K is flexibility, negative to N and positive to self-integrating resistance R. Therefore, the voltage $u(t)$ is linear with monitored current $i_1(t)$. In practice, the integral resistance R is usually connected in parallel with a stray capacitance C_0, such as the signal cable connected to the output end. We then have the differential equation

$$e(t) = LC_0 \frac{d^2u(t)}{dt} + \left(\frac{L}{R} + R_L C_0\right)\frac{du(t)}{dt} + \left(1 + \frac{R_L}{C_0}\right)u(t) \tag{6.20}$$

Putting equations (6.16) and (6.20) into a Laplace transform and setting the initial condition to zero, we have the transfer function as follows:

$$H(s) = \frac{u(s)}{I_1(s)} = \frac{R}{N} \cdot \frac{s}{RC_0 s^2 + \left(1 + \frac{R_L C_0 R}{L}\right)s + \frac{R_L + R}{L}} \tag{6.21}$$

For a self-integral broadband sensor $R_L C_0 R/L \ll 1$, so

$$H(s) = \frac{R}{N} \cdot \frac{s}{RC_0 s^2 + s + \frac{R_L + R}{L}} \tag{6.22}$$

Taking the module from equation (6.22), we have the following characteristics of amplitude–frequency:

$$H(\omega) = |H(j\omega)| = \frac{1}{C_0 N} \frac{\omega}{\sqrt{\left(\frac{R_L + R}{RC_0 L} - \omega^2\right)^2 + \left(\frac{\omega}{RC_0}\right)^2}} \tag{6.23}$$

When $\omega = \omega_0 = \sqrt{\frac{R_L + R}{RC_0 L}}$ $|H(j\omega)|$ reaches its peak, we have

$$H(\omega)_{max} = |H(j\omega)|_{max} = K = \frac{R}{N} \tag{6.24}$$

The result is equal to (6.19), so we have

$$f = f_0 = \frac{1}{2\pi}\sqrt{\frac{R_L + R}{RC_0 L}} \tag{6.25}$$

Usually $R_L \ll R$, equation (6.25) changes to

$$f_0 = \frac{1}{2\pi\sqrt{LC_0}} \tag{6.26}$$

Here, f_0 refers to the resonant frequency of the sensor, such as 3-dB broadband – i.e., $H(\omega) = \dfrac{1}{\sqrt{2}}|H(j\omega)|_{\max}$. Estimating upper and lower frequency limits, ω_H and ω_L, we have

$$\omega_H \omega_L = \omega_0^2 = \frac{R_L + R}{R C_0 L} \tag{6.27}$$

For broadband

$$\omega_H - \omega_L = \Delta\omega = \frac{1}{R C_0} \tag{6.28}$$

where ω_H is bigger in practice than ω_L by an order of magnitude or above, so we have

$$\begin{cases} \omega_H \approx \dfrac{1}{R C_0} \\ f_H = \dfrac{1}{2\pi R C_0} \end{cases} \tag{6.29}$$

with

$$\begin{cases} \omega_L \approx \dfrac{R + R_L}{L} \\ f_L = \dfrac{R + R_L}{2\pi L} \approx \dfrac{R}{2\pi L} \end{cases} \tag{6.30}$$

Broadband sensors are designed in accordance with equations (6.16), (6.24), and (6.30). Table 6.9 lists the measured results of a ferrite magnetic core's influence on the sensor characteristics (sensitivity K) with different numbers of turns N and integral resistance R. As you can see, K is directly proportional to R, but inversely proportional to N. f_L increases as R increases, while f_H reduces as R and C_0 reduce and as N increases. When the sensor transports through 20 m of transmission cable, the upper limit of the frequency f_H reduces by an order of magnitude as C_0 increases.

Table 6.9 Characteristics and parameters of broadband sensor

		Measured results			Through 20 m of cable		
N	R (kΩ)	f_L (kHz)	f_H (kHz)	K (V.A^{-1})	f_L (kHz)	f_H (kHz)	K (V.A^{-1})
50	2.50	39.0	530.0	48.6	22.0	52.0	47.0
50	1.20	17.8	923.0	24.0	18.2	77.0	23.0
50	0.62	7.1	1650.0	12.3	7.4	138.0	12.0
50	0.31	3.5	2000.0	6.2	3.5	272.0	6.2
25	0.62	30.0	1622.0	24.4	30.0	149.0	24.0
25	0.31	14.0	2050.0	12.3	14.0	259.0	12.0
25	0.15	7.0	2064.0	6.0	7.0	589.0	6.0

6.1.1.4.4 Principles for Arrester Condition Monitoring

Today, China's MOA condition-based monitoring methods include leakage full-current, resistive current harmonic analysis, resistive current third harmonic analysis, and compensation. In this world, dual-AT and temperature-based measurement methods are developed.

Method of Full-Current Leakage When the MOA ages or gets damp, the component of the resistive current increases and subsequently the full current will increase. Operators can make a judgment on MOA conditions by use of this characteristic. The method for monitoring full current can follow the principle shown in Figure 6.10. Under the same grid voltage, the grain boundary capacitance C is approximately constant, so there is no obvious change in capacitive component. Hence, an increase in full current is caused by the increased resistance component. It is therefore an easy operation for users to monitor the change in full current, since it suggests a change in resistance current to some degree. However, when the MOA is normally operated (i.e., when the MOA is neither aging nor affected by damp), the resistance component of the full current has a paltry 10% capacitive component and the difference in fundamental wave phase reaches 90°. This means that the effective value or mean value of the full current depends mainly on the capacitive current component. Even though the resistance current increases multiple times, the change in full current is not obvious. When the peak value of the resistance current increases from 1 to 4 μA, the possibility that the full current increases is just a few percent. There is little sensitivity to leakage current by monitoring the MOA full current. It can tell an obvious change only when the device is affected by damp or

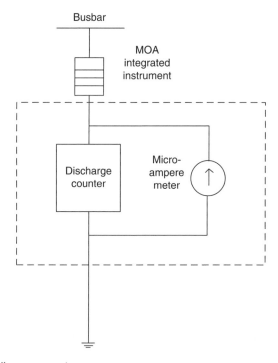

Figure 6.10 MOA full current monitor.

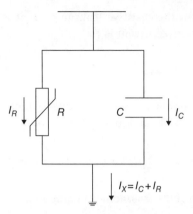

Figure 6.11 Equivalent circuit of MOA valve block under small current.

aging, which is not the best way to find an early fault in the MOA. Monitoring the full current through fiber-optical sampling could solve problems such as HV isolation and data transmission over a long distance, and can be used to monitor the leakage current and alarm in time, but it still does not help to improve the sensitivity.

Harmonic Analysis for Resistance Current Figure 6.11 shows the model for an equivalent circuit of the MOA valve block at a single-phase small current and with the circuit connected in parallel by nonlinear and linear resistors. In the equivalent circuit, the total leakage current through the MOA is divided into a resistive current and a capacitive current. The reactive loss produced by the latter does not give out any heat to the valve, instead, the valve is heated by the active loss produced by the resistive current. Assume U_X to be the working voltage, I_X the total leakage current, I_R the resistive current, and I_C the capacitive current. The power system voltage U_X or current I_X that meet the Dirichlet conditions can be decomposed by Fourier series into

$$U_X = U_0 + \sum_{k=1}^{\infty} U_{km} \sin(k\omega t + \alpha_k) \tag{6.31}$$

$$I_X = I_0 + + \sum_{k=1}^{\infty} I_{km} \sin(k\omega t + \beta_k) \tag{6.32}$$

where U_0 is the DC component of voltage, I_0 is the DC component of current, U_{km} is the harmonic amplitude of voltage, I_{km} is the harmonic amplitude of current, α_k is the phase angle of voltage, and β_k is the harmonic wave phase ($k = 1, 2, 3, 4, ...$).

As we can see from Figure 6.11:

$$I_X = I_C + I_R \tag{6.33}$$

We have the capacitive current according to equation (6.31):

$$I_C = C \frac{dU_X}{dt} = \sum_{k=1}^{\infty} I_{Ck} \cos(k\omega t + \alpha_k) \tag{6.34}$$

where $I_{Ck} = kC\omega U_{km}$ is the harmonic amplitude of the capacitive current.

Since the phase angle of the resistive leakage current is equal to that of the kth harmonic voltage, the resistive current is

$$I_R = I_0 + \sum_{k=1}^{\infty} I_{Rk} \sin\left(k\omega t + \alpha_k\right) \tag{6.35}$$

where I_{Rk} is the harmonic amplitude of the resistive current.

Taking equations (6.32), (6.34), and (6.35) in equation (6.33), we have

$$I_0 + \sum_{k=1}^{\infty} I_{Rk} \sin\left(k\omega t + \beta_k\right) = \sum_{k=1}^{\infty} I_{Ck} \cos\left(k\omega t + \alpha_k\right) + I_0 + \sum_{k=1}^{\infty} I_{Rk} \sin\left(k\omega t + \alpha_k\right) \tag{6.36}$$

According to the nature of the trigonometric function, we have

$$I_{Rk} = I_{km}\left(\cos\alpha_k \cos\beta_k + \sin\alpha_k \sin\beta_k\right) \tag{6.37}$$

$$I_{Ck} = I_{km}\left(\cos\alpha_k + \cos\beta_k + \sin\alpha_k + \sin\beta_k\right) \tag{6.38}$$

Taking equations (6.37) and (6.38) in equations (6.34) and (6.35), respectively, we have expressions for resistive and capacitive harmonic current. In this way, the total resistive leakage current of the MOA can be obtained.

Generally, one factor that leads to the degradation of arrestor insulation is zinc-oxide valve aging, which worsens the performance of nonlinear characteristics. For instance, the high-order harmonic component of the resistive current increases greatly, while the fundamental component increases only slightly. The other factor is being affected by damp. For instance, the fundamental component of the resistive current under normal voltage increases significantly, while the high-order harmonic component increases only slightly. Therefore, through the measurement of each harmonic of the resistive current, operators can identify the reason why the performance of a zinc-oxide MOA deteriorates.

Method of Third Harmonic for Resistive Current Since the MOA is highly nonlinear, the resistive component of the full current includes not only the fundamental wave, but also third, fifth, and higher-ordered harmonics. Gradually, the fundamental wave becomes less and less. In the MOA leakage current, the third harmonic I_{r3} is a characteristic quantity that is sensitive to aging and failures. The third harmonic method for resistive current is to detect the third harmonic component with a bandpass filter; the peak of the resistive current can be obtained according to a certain relation between the MOA resistive current and the third harmonic resistive component. The method of monitoring the third harmonic resistive current from the three-phase total current is also called the method of zero-sequence current. The principle of monitoring is shown in Figure 6.12. The third harmonic resistive current can be measured through the small TA via a three-phase grounding wire. At the same time, the resistive current I_R can be obtained according to the calculated relation between the three-harmonic resistive current I_{r3} and the resistive current. This is easy, but it has shortcomings. For instance, the busbar voltage includes a certain proportion of harmonic voltage and the capacitive harmonic current that comes with it will have an effect on the measurement results. Also, when different types of MOA are aging, the high harmonic component of the resistive current follows different rules, so the degree of aging of the MOA is hard to judge by a unified criterion.

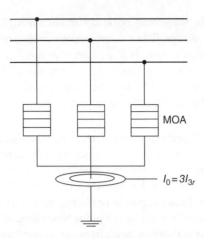

$I_0 = 3I_{3r}$

Figure 6.12 Principle of measuring MOA resistive three-harmonic current.

Capacitive Current Compensation　This refers to obtaining the resistive current I_R by compensating for I_C in the full current I_X. The calculation principle can be expressed with the formula

$$\int_0^{2\pi} U_{s0}(I_X - GU_{s0})d(\omega t) = 0 \tag{6.39}$$

In equation (6.39), U_{s0} is the result of a phase shift by 90° from the external voltage – i.e., in phase with the capacitive current I_C. When fully compensated $(I_X - GU_{s0})$, the capacitive current is equal to I_R. According to equation (6.39), we have a compensation coefficient G. Then, we have the component of the resistive current with formula

$$I_R = I_X - I_C = I_X - GU_{s0} \tag{6.40}$$

Following this principle, the LCD-4 leakage current measuring equipment has been China's most commonly used instrument so far. It applies the voltage with a normal waveform on a single MOA, so the measured resistive current is accurate. But there are many problems during three-phase operation, since we cannot ensure that we can remove the phase-to-phase coupling current, and obtain a reasonable zinc-oxide arrester resistive current correctly by the method. Errors that may occur in the actual measurement by the method of capacitive current compensation include the following.

1) Three-phase arresters are installed in a "-" pattern. Owing to the interference of phase-to-phase coupling capacitance and EMI, the arresters of each phase are affected not only by the phase voltage, but also by the nearby voltage of phase-to-phase coupling. Thus, the accuracy of the result is affected and the result shows that A is bigger than B and B is bigger than C. For a 110 kV system, the error in phases A and C is about 10%, and the error in a 330 kV system reaches 25%. Systems with a higher voltage grade will have a bigger error.

2) In the nonlinear branch current of the zinc-oxide valve, the AC volt–ampere characteristic curve shows hysteresis in varying degrees when the current or voltage crosses a zero point. This indicates that when the network voltage is in the waveform of a

sinusoidal function, the waveform peak in current through the MOA is not coincident with that of the voltage. The current waveform is presented in the form of an odd harmonic function. So, there is a large error in the measured resistance current; errors in the fundamental wave of the current in a nonlinear branch circuit reach 15%.

3) When the network voltage contains a harmonic, it cannot remove the capacitive harmonic current and thus may lead to harmonic current errors.

6.1.1.4.5 MOA Condition-Based Monitoring Technology Outside China

Method of Dual AT   One AT sensor uses a normal leakage current and the other impacts the peak of a large current at over-voltage to record the times of MOA action. MOAs are distinguished by 2.5 kA and 20 kA reference current (e.g., lightning strike or over-voltage), and go through digital signal processing after A/D conversion. The voltage signal received by an optical fiber is used to judge how the grid harmonic influences the resistive component of the leakage current. In order to confirm that the leakage current is not increased by temperature, a temperature sensor is installed to measure the ambient temperature in the vicinity of an MOA. The principle is shown in Figure 6.13.

A dual AT relies on powerful software to make condition-based monitoring possible. At the same time, influences of the power grid harmonic and temperature are considered. It is better than the existing methods in terms of functions, but is not that economical. A HV MOA has a service life of 20 years. It remains to be seen whether this method will stand the test of time as regards stability in the long term.

Temperature-Based Measurement Method   This method follows the principle that when an MOA ages or gets damp, the temperature will increase as the leakage current increases.

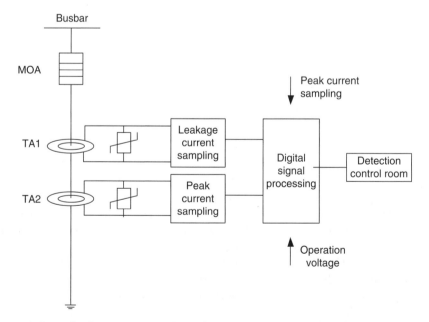

Figure 6.13 Principle of monitoring MOA by dual AT.

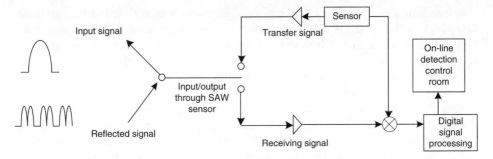

Figure 6.14 Condition-based monitoring system based on temperature measurement.

In case of over-voltage, the temperature may rise temporarily, but would return to normal gradually. The measured temperature is not a direct way to know the state of operation. Temperature can affect the MOA operation state parameter comprehensively due to a continuous operating voltage, with overheating of the MOA directly related to energy loss and not directly related to quality of voltage or external disturbance. A temperature sensor, if installed in the arrester, makes it hard to enclose the arrester. Moreover, the temperature sensor will not suffice to get signals from the whole arrester. The SAW temperature sensor, developed in Germany, requires no power. The state-monitoring system follows the principle shown in Figure 6.14. With high-frequency signals from the oscillator (at 30 MHz to 3 GHz), the SAW sensor receives the signal via a valve plate and reflects a signal with temperature. The site receiving equipment collects a high-frequency signal, which is processed via a digital signal temperature waveform after reference to the ambient temperature.

A passive SAW sensor is generally made into a valve-like shape and installed in the middle of the valve plates of the central MOA. It sends and receives HF signals, which are almost immune to site interference. In heavily polluted areas, the contaminated leakage current on the VOA surface may lead to overheating, so the operators will take anti-pollution measures to protect the local area. This method makes a difference to MOAs that are under manufacture and ready to be installed with condition-based monitoring. It does not apply to MOAs in service.

6.1.1.5 Capacitive Equipment Status-Detection System
6.1.1.5.1 Capacitive Equipment Failures
Among all high-voltage electric power equipment in an electric system, the capacitive equipment takes up a large proportion of the insulation structure, including current sensor, bushing, coupling capacitor, capacitive potential transformer, etc. Therefore, its operation will exert a direct influence on the safe, stable, and economical operation of the electric system. With increases in high-voltage level and capacity, it takes more time and costs more in maintenance/correction than traditional time-based methods, which obviously fail to meet the requirements for continuous development of modern smart power grids. It is a trend for electric systems to use condition-based maintenance instead of time-based maintenance. As a basis for condition-based maintenance, on-line maintenance will help save time, cut costs, and make service life longer. At the same time, this method may be able to predicate and remove problems to maintain normal operation, improve the reliability of the power grid, and ensure the safety of people and equipment.

Research on the technology of detecting insulation status for capacitive equipment dates back to the 1960s. For decades, many products with advanced technology have been put into operation, such as the insulation status-detection system for capacitive equipment developed by the former Soviet Union; the monitoring unit of tan δ for capacitive equipment by Australia; and other on-line units for capacitive devices by companies in America and Canada.

Since the 1980s, Chinese institutions and manufacturers have taken up research on monitoring the status of the general characteristic parameters of insulation for transmission and transformation equipment (including transformer, mutual transformer, and coupling capacitance). The characteristic parameters include the current passing through the equipment insulation, capacitance, dielectric loss factor tan δ, etc. Years of unremitting effort have led to significant achievements, including some status-monitoring units that have good performance in the field.

As the sensor, communication, and computer technologies develop and are applied in many areas, there are increases in status-monitoring parameters and methods; partial discharge, infrared temperature measurement, gas chromatography, and ultrasonic location have all been applied to insulation-status monitoring. Many other research achievements and monitoring devices have gained a lot from the field of application as well.

The dielectric loss factor tan δ plays an important role in the on-line status monitoring of capacitive equipment, and therefore has been given top priority in research. It focuses on principle methods, status-monitoring sensors, signal processing and analysis methods, and research into practical applications.

Compared with traditional commonly used measurement methods, the tan δ monitoring method has its own advantages, however, there are still difficulties in practical application. The δ angle is very small. If tan $\delta = 0.1$, then $\delta = 0.06°$ and it is not an easy job to measure such a small phase position. What's more, on-line measures suffer interference by many factors.

1) *It is difficult to receive real signals.* A weak current passes through the ground wire on electrical equipment. What's more, the requirement for measurement precision is high. Plus, given the interference of the electromagnetic field, it is difficult for micro-current sensors to keep high measurement precision over the long term.

2) *Strong interference from site.* The equipment is installed in complicated environments, where there is strong interference. Therefore, it is hard to acquire a true signal, especially when strong interference caused by the switching operations of capacitive equipment not only affects the measurement results but also damages the measuring equipment.

3) *Work frequency power grid fluctuation.* Harmonic analysis works in such a way that tan δ is determined by a fast Fourier transform (FFT) algorithm. The fact that the pre-sampling frequency is out of sync with the working frequency of the power network due to the latter's instability may lead to spectrum leakage and a large error in tan δ.

6.1.1.5.2 *Principle for Dielectric Loss Factor Condition Monitoring*

Dielectric (insulator) loss makes reference to the energy loss during the conversion to thermal energy from dielectric under the effect of an alternating electric field. Great loss may lead to a rise in temperature and even aging, such as a tendency to be fragile and decompose. When the temperature rises to a certain value, there is a risk of thermal runaway for the dielectric. Besides, the electrical equipment is susceptible to environmental factors (such as oxidization, light, humidity, chemical substances, and microorganisms) that cause irreversible phenomena (such as damage to the internal/external structure of the dielectric and poor performance). More seriously, insulation failure may occur. Insulation materials can scarcely conduct electricity at DC voltage, except for current leakage on very small surfaces. While current passes through the insulation materials at AC voltage, therefore, the insulation materials at AC voltage are considered equally as a shunt or serial circuit by capacitive and electric resistance. Figure 6.15 shows the equivalent circuit of capacitive equipment at AC voltage and a phasor graph for the voltage and current in the circuit.

Taking the dielectric shunt equivalent circuit as example, the resistance of the equivalent shunt circuit tends to infinity when the electrical equipment has good insulating materials (no dielectric loss), therefore, the loss can be ignored for an AC power plant and the total current I passing through the electrical equipment is the pure capacitive current I_C whose phase is ahead of the voltage phase by $\pi/2$. When the insulating materials get worse (with dielectric loss), the resistance of the equivalent shunt circuit will be reduced and the current I passing through the insulation media contains the resistive current I_R. So, the current advances the voltage by less than $\pi/2$. $\tan \delta$, as the ratio between resistive current and capacitive current, is a reflection of dielectric loss – that is, $\tan \delta = I_R/I_C$. Further, $\tan \delta$ can be derived from the formula for loss angle tangent or dielectric loss factor:

$$\tan \delta = \frac{I_R}{I_C} = \frac{P}{U^2 \omega C} = \frac{1}{R\omega C} = \frac{1}{2\pi f\left(\varepsilon \dfrac{s}{d}\right)*\left(\rho \dfrac{d}{s}\right)} = \frac{1}{2\pi f \varepsilon \rho} \tag{6.41}$$

where ω is the angular frequency, ρ is the resistivity of the insulating medium, and ε is the dielectric constant.

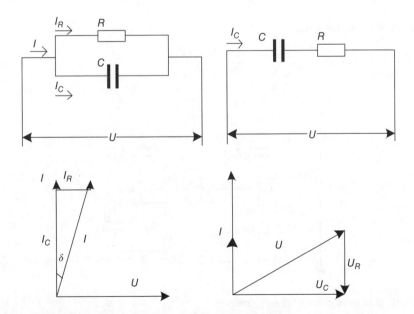

Figure 6.15 Equivalent circuit and phasor graph for lossy dielectric.

As we can see from equation (6.41), tan δ is related to the power frequency, dielectric constant, and resistivity, and has nothing to do with the dielectric volume (this factor actually reflects the dielectric loss per unit volume), so it can reflect the distribution faults of the dielectric insulation and therefore the monitoring of the capacitive equipment insulation focuses on the characteristic parameter tan δ.

6.1.1.5.3 Measurement of Dielectric Factor tan δ

A bridge method is traditionally used and the Schering bridge is a typical method which calculates the C_X and tan δ characteristic parameters of a dielectric for capacitive equipment through bridge balancing by comparing AC bridge differences. The Schering bridge circuit structure includes a bridge body, standard capacitor, and test power source. A schematic diagram is shown in Figure 6.16.

If the bridge is balanced, the galvanometer G will show $I_G = 0$, i.e.

$$I_{CE} = I_{AC} = I_X$$

$$I_{DE} = I_{AD} = I_N$$

$$I_{CE} = I_{DE}$$

$$U_{AD} = U_{AC} = U_X$$

The complex impedance of each bridge arm should satisfy the formula

$$Z_3 Z_N = Z_4 Z_X \qquad (6.42)$$

where Z_X is the equivalent impedance of the test dielectric, Z_4 is the equivalent parallel complex impedance between R_4 and C_4. Taking

$$Z_3 = R_3, \; Z_N = 1/j\omega C_N, \; Z_4 = \frac{1}{\dfrac{1}{R_4} + j\omega C_4}, \; Z_X = \frac{1}{\dfrac{1}{R_X} + j\omega C_X}$$

in equation (6.42), we have

$$\left(\frac{1}{R_X R_4} - \omega^2 C_X C_4 \right) + j\left(\frac{\omega C_4}{R_X} + \frac{\omega C_X}{R4} \right) = j\frac{\omega C_X}{R_3} \qquad (6.43)$$

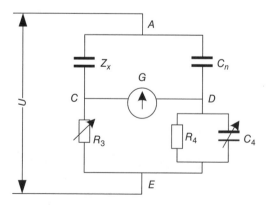

Figure 6.16 Wiring diagram for Schering bridge.

Since the real part is equal to the imaginary part in equation (6.42), we have

$$\tan \delta = \frac{1}{R_X \omega C_X} = \omega R_4 C_4 \tag{6.44}$$

$$C_X = \frac{C_N R_4}{R_3} \cdot \frac{1}{1+\tan\delta} = \frac{C_N R_4}{R_3} \quad (\text{in pF}) \tag{6.45}$$

We consume $R_4 = \dfrac{10^4}{\pi} \approx 3184\,(\Omega)$. If the power frequency is $50\,\text{Hz}$, $\omega = 2\pi f = 100\pi$, then we have

$$\tan\delta = 2\pi f \cdot \frac{10^4}{\pi} = 10^6 C_4 \quad \left(C_4 \text{ is in F}\right)$$

$$\tan\delta = C_4 \quad \left(C_4 \text{ is in } \mu\text{F}\right)$$

C_4 is representative of the adjustable capacitance until, at balance, the value of C_4 is $\tan\delta$ of the tested product.

The dielectric loss factor $\tan\delta$ is traditionally measured by means of a Shering bridge when it is powered off. This method enjoys high precision, since the bridge cannot keep balanced and causes a large error in the event of significant harmonic interference or a strong electric field. The Schering bridge applies in places with small interference, including the lab. Many documents have proposed solutions to interference, such as shielding, power-phase inversion, THG test power, graded compression, substitution, anti-interference, and interference with power supply. Despite all of this, the solutions have their own defects and are not all applicable in all fields. In the Shering bridge method, a HV standard capacitor is used or a through-voltage transformer secondary voltage and piezoelectric capacitor to measure the value of tan 0. However, owing to the high costs, transportation, complex test program, low level of automation, heavy loads, and human factors, it is seldom used in this field.

6.1.1.5.4 *Modern Measurement Method*
To overcome the difficulties caused by limitations existing in traditional methods for measuring the value of $\tan\delta$ in many aspects, we are desperately seeking for a better approach to monitoring insulation materials. Long-term research proves that capacitive equipment insulation supervision is the technology to be researched and developed for smart power grids. There are basically two ways to supervise the insulation conditions of capacitive equipment: one is through hardware where the phase angle is measured directly, such as the zero-point crossing measurement method or voltage comparator; the other is to generate a quantity by software and have it number signal processed. In so doing, we obtain $\tan\delta$.

Method of Crossing-Zero Phase Comparison This method belongs to the hardware approach that is commonly used at home and abroad for on-line measurement of the dielectric loss factor. We obtain $\tan\delta$ through the sinusoidal current of the capacitive equipment and the phase difference at a zero-crossing point measured by impulse counting. The current and voltage signals of the capacitive equipment are passed through a zero-crossing comparator and consistent square signals are generated. After reading the difference between power and voltage signals with CPU impulse counting, we get the phase difference. The block diagram for such a measurement method is shown in Figure 6.17.

From this figure we see how the insulation monitoring units on capacitive devices measure the dielectric loss factor (tan δ) by way of zero-crossing phase comparison. A micro-current sensor connected to the grounding wire of the capacitive equipment C_x is designed to measure the micro-current of the capacitive equipment to ground I. A potential transformer at the bus side collects the voltage to ground. The mutual inductor secondary voltage will drop to the expected nominal voltage U at a certain turn ratio. According to the measured current and voltage, another insulation characteristic parameter (capacitance) can be calculated from the equation $C = I/U$. The angle differences between micro-current sensor and potential transformer are ignored and the phase difference between U and I is $90° - \delta$, where δ represents the dielectric loss angle.

The time sequence for a zero-crossing point phase comparison is shown in Figure 6.18. In this figure, U_u expresses the standard voltage generated by transference of voltage

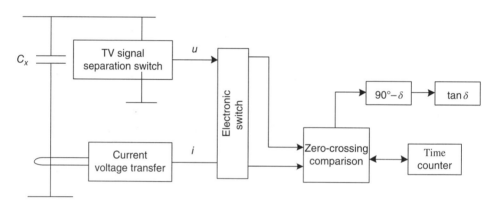

Figure 6.17 Schematic diagram of measurement method.

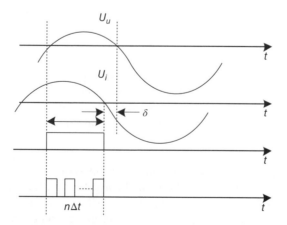

Figure 6.18 Zero-phase sequence of time by comparison.

signals and U_i expresses the voltage generated by transference of current signals. Flowing through a zero-comparison circuit, it will generate SW sync signals correspondingly. The time difference is measured at zero crossing (rising/falling edge) – i.e., square wave width T $(T = 90° - \delta)$. The square wave width, expressed as T, is read with a HF clock pulse. With number of HF clock pulses n and clock period t, we can calculate the phase angle δ according to the formula $T = n\Delta t$.

Assume that the dielectric loss is equal to 0.06°, then $\delta = 0.06°$, $\tan\delta = 1$, the signal frequency is 50 Hz, the square wave width T is equal to 0.00499(s) $\left[T = \dfrac{90° - \delta}{50 \times 360°} = 0.00499(s) \right]$, the clock period t is equal to the clock square wave, so the number of monitoring clock waves is

$$n = T / \Delta t = \frac{0.00499 \times 10^6}{0.5} = 9980$$

As we can see, the method measures the zero-crossing point (ZCP) time difference and sinusoidal signal cycle by taking advantage of CPU counting impulses. It is simple in terms of design, moreover, there is a way to get a more accurate $\tan\delta$ by measuring several sets of data, and taking an average value after removing the maximum and minimum values to make it less scattered and reduce errors. This is, therefore, one of the most mature technologies. Considering that the δ angle is small, when $\tan\delta = 0.1$, $\delta = 0.06°$. For a 50 Hz AC signal, this amounts to 3 µm pulse width, making it difficult for the CPU to count and measure. In addition, field interference may affect the measurement accuracy. Therefore, it is worthwhile carefully studying how to improve the accuracy of the measurement and how to reduce the impact of interference on the accuracy of the measurement. Currently, to improve the precision of the measurement, China has focused on changing the measuring units that use the method of ZCP comparison and processing the measured data – such as pre-test self-correction, zero-phase shifting filter, and taking the mean value after dual-direction ZCP voltage comparison.

Method of Harmonic Analysis   As one of the most used methods by the insulating monitoring software of capacitive devices, this works as follows. First, magnify the current collected by the current sensor connected to the grounding wire of the capacitive equipment, and the standard voltage signals from the potential transformer secondary side via a front-end amplifying circuit. Second, with the conditions of A/D conversion satisfied, pass these signals through an A/D conversion circuit to convert analog signals into digital signals. Third, put the dispersed digital signal waveform into an FFT with the use of a processor to get the fundamental Fourier coefficients, respectively, and the phase difference between the two fundamental waves. Thus, $\tan\delta$ is calculated. Assume U_X is the working voltage at the bus side of the capacitive equipment and I_X is the current through the equipment, so we can further break this up by Fourier series:

$$U_X = U_0 + \sum_{k=1}^{x} U_{km} \sin(k\omega t + \alpha_k)$$
$$I_X = I_0 + \sum_{k=1}^{x} I_{km} \sin(k\beta t + \beta_k)$$

(6.46)

where U_0/I_0 is the DC component of the voltage/current, respectively; U_{km}/I_{km} is the harmonic component of the voltage/current, respectively; α_k/β_k is the harmonic phase angle of the voltage/current, respectively. So, the dielectric loss factor is $\tan\delta = \tan[90° - (\beta_1 - \alpha_1)]$.

According to the above analysis of measurement methods, we know that the software approach to on-line measurement factors out the fundamental wave phase angles from the current and voltage signals of the tested goods, which are rapidly transferred using FFT, and finally the phase angle δ is calculated. The FFT algorithm is mainly used to calculate $\tan\delta$ in the method of harmonic analysis. Therefore, frequency fluctuation in the work frequency grid would result in a pre-sampling frequency out of sync with the work frequency, as well as leakage that may lead to errors in $\tan\delta$.

Despite the advantage of high precision in the Schering bridge method, harsh environment and lack of on-line monitoring are other problems. The method of zero-point cross phase comparison has higher requirements for hardware, to begin with. For example, it should have high precision, very stable micro-current sensor and signal processing unit, and high resistance to interference. So, it is rather difficult to make stable and precise measurements. One typical feature of harmonic wave analysis is that it is based on the Fourier transform. Based on the analysis results of current and standard voltage signals, it can find the phase angle of each harmonic wave using FFT arithmetic and further calculate the dielectric loss factor $\tan\delta$, keeping the result free from the influence of the upper harmonic at the same time. In addition, according to the characteristic of orthogonality for a trigonometric function, the DC signals U_0/I_0 in U_X/I_X are not affected by A1 and B1, that is to say, the method of harmonic wave analysis is immune to circuit zero drift, which makes measurement more stable and precise. Since the coefficient is mainly used for software analysis, it may reduce the influence of the hardware circuit and also change the practice of solving problems only by improving the hardware, which greatly increases the costs. Work frequency fluctuation will affect the measurement precision.

6.1.1.5.5 Sensor

The sensor is a key element used for on-line measurement of the dielectric loss factor $\tan\delta$, and its measurement precision will make a great difference to that of $\tan\delta$. Therefore, sensor performance takes on an important role in measurement of the monitoring system, and high quality has been given top priority for accurate measurement.

Micro-current sensors, generally located on the ground wire of the capacitive equipment, is a key element for measuring the dielectric loss factor, and mainly used to acquire original signals in the on-line monitoring system. There is a weak current passing through the micro-current sensor when the capacitive equipment works in normal conditions, varying from hundreds of microamperes to milliamperes. Therefore, the original signals received are weak. In addition, poor field conditions and intensive electromagnetic interference also make it more difficult to collect original signals. All these factors set higher requirements for the performance of micro-current sensors, and they should meet the conditions listed below.

1) Sensors should be provided with electrical isolation between the primary-side capacitive equipment and the secondary equipment measuring device to keep the measuring device on the secondary side safe in case of failures in the capacitive equipment on the primary side.

2) The acquisition value should have high precision and strong anti-interference, and be able to convert to the signal as required.
3) It should be easy to install and commission, with good properties and economy/durability from an economic point of view.

A micro-current sensor is a special kind of transformer. The primary and secondary windings are coupled to each other via a magnetic flux. Figures 6.19 and 6.20 show schematic diagrams of a T-shaped equivalent circuit and sensor access methods, respectively.

In Figure 6.19, I_1/I_{2N} are the current converted into the secondary side from the primary/secondary current; Z_1 is equal to the circuit resistance, including leakage reactance and winding resistance; Z_{2N} refers to the equivalent value of the secondary circuit resistance, including the leakage reactance and internal resistance of secondary winding; Z_{1N} refers to the value from the secondary load impedance; Z_{1m} is equal to the excitation winding impedance; I_0 is equal to the excitation current. The micro-current sensor ratio error f is equal to the relative error between the secondary current transferred to the primary side by the rated transformation ratio and the actual secondary current, calculated according to the formula

$$f = \left[(K_n I_2 - I_1)/I_1 \right]\%$$

(6.47)

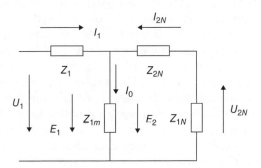

Figure 6.19 Equivalent circuit of current sensor.

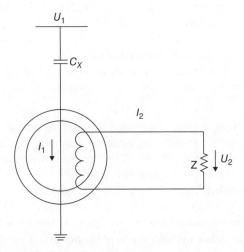

Figure 6.20 Access method of sensor.

The phase-angle difference for the current sensor refers to the difference between the reversed secondary current and the actual primary current. It is theoretically considered that the excitation current I_0 is one factor leading to an error in sensor. As seen from Figure 6.19, since $I_0 \neq 0$, $I_1 \neq -I_2 N$, so the ratio and angle difference are generated; if $I_0 = 0$ and there is no excitation current, we can say that the current sensors are working in good conditions where there is a difference of 180° between the primary and secondary current and no angle or ratio difference.

The sensor magnetizing current is never the sole reason for ratio differences and angular differences from an engineering aspect. Sensor primary and secondary windings leakage, high nonlinearity of iron-core soft magnetic materials, and lack of stability are important factors for ratio error and angular differences. What's more, micro-current sensors are installed outdoors, which requires a good temperature characteristic as well as mechanical properties, while the property relating to angle differences of micro-current sensors is subject to temperature.

6.2 Distribution Network Device Monitoring System and Network Monitoring Management System

The distribution network device monitoring system varies from substation, to feeder line, to regional transformer. The substation is equipped with both a distribution substation and a switching station; the distribution line includes both overhead line and underground cable; customer substations are of various kinds. Because of the wide variety of devices and scattered layout, the device monitoring system for a distribution network should not be identical with that of a substation; instead, it should be fully reliant on the existing monitoring management system for the distribution network.

At present, the distribution network monitoring system should be constructed in a hierarchical way to improve the speed and meet the requirements for openness and expandability. The monitoring and management system for a distribution network includes the terminal, substation, and main station layers.

1) The terminal layer mainly includes the monitoring control units, such as FTUs and TTUs. The monitoring units include the transformer, capacitive equipment, breaker, feeder, and regional transformer.
2) The substation layer consists of the server, workstation, and other auxiliary equipment connected via a communication system.
3) The communication layer accurately sends commands from the main station to the terminals, and saves information about the conditions of the remote equipment in the main station.

The power transmission and transformation monitoring systems are constructed in a similar way, and divided into monitoring terminal, communication network, and main station monitoring management software. The structure is similar to that of the device monitoring system for the distribution network.

It is unrealistic to establish an independent supervisory and maintenance system due to the strict controls on construction, operation, and maintenance costs of the distribution network and distribution network equipment. Since the present distribution management system is rather mature, the supervisory system can rely on the existing automation device, communication channel, and system software to avoid repeated construction and save costs.

6.2.1 Distribution Network Equipment Supervisory Terminal and Distribution Network System Terminal Layer

The distribution network auto-remote terminal unit is widely used in 10-kV distribution lines, and supervises the circuit breaker, load switch, looped contact switch, and reactive compensation capacitor complete with communication part communicating with the SCADA master station. In so doing, all the distribution lines are monitored.

Taking the feeder supervisory control terminal for example, the FTU controls the feeder and switchgear in the supervisory management system. General functions of the FTU include the following.

1) As a rule, the measurement and off-limits monitoring of current is realized through the acquisition of analog information to determine whether it is normal or faulty current, and of AC input voltage to monitor the supply of power to the feeders on the ends of the switchgear.
2) Condition values, reports, and records of state quantities and times are collected, then the remote control command is received and acted on.
3) *Block function.* This operates the switch device according to the remote control command and automatic control.
4) *Communications.* Communication takes place with the upper master station, reporting the collected message about acquisition and receiving/processing the control command from the upper station, with a local communication interface.
5) *Message storage.* This maintains power when the power is off; the recorded data can be saved for a long time.
6) *Maintain working when power off.* The device is prepared with backup power, which maintains working for over 24 h when the line and main power source lose power; it has the function of an operating switchgear.
7) *Device configuration function.* The device can be configured with the operating mode of the terminal and protection set value.

These characteristics make the FTU an important monitoring control unit and link in the automation system of distribution automation.

As the terminal device in the digital acquisition system, the FTU features acquisition, processing, and control of analog/digital values and capacity of communication, and is fully competent to construct monitoring functions for the circuit breaker and ring main unit using the characteristics of expansibility so as to construct a terminal layer for monitoring distribution devices.

When the FTU is designed as a terminal of the monitoring switchgear, in respect of the monitoring switchgear, besides the functions above, the FTU should have the following extra functions.

1) *Recording, calculating, and storing messages about the real-time conditions of devices.* Each monitoring unit is supposed to collect a short-circuit breaking current, load breaking current, and phase voltage, and calculate accumulative values. Under normal conditions, the three-phase load current flowing through the circuit breaker is recorded and calculated so that these data can be used by maintenance personnel

inspecting the switchgear and device. For example, it records the action time, phase, and breaking current in the operation of breaking or closing so as to recognize the behavior of the action when the circuit breaks permanently. In case of fault, it collects or records the current in the open circuit or short circuit so as to provide an analysis program for the waveform and harmonic wave analysis. Performance permitting, the terminal is able to collect electrical and mechanical parameters of the circuit breaker during operation.

2) *Recording statistics of historic information about equipment operation and test.* Distinguishing debugging opening, separating brake of operation and fault, which records the accumulative opening times and total times of three different types of opening. It records the moment when the switching device starts or stops and the accumulative outage time of switch devices.

3) *Saving the status of conditions.* The recorded value will be kept for a period of time after power loss. The use of an FTU installed with or without the expanded functions makes no difference to distribution system building. Totally dispersedly located, it is later placed in the terminal box of the switchgear inside the battery limit by monitoring devices. In a similar way, the TTU and DTU terminals are integrated with the function of monitoring conditions. Relying on the terminal layer of the automation system, it is possible to build a terminal layer for monitoring conditions with a small amount of funds.

6.2.2 Condition Monitoring System Relies on Automation System Communication Channel

The special structure of the distribution network determines the characteristics, such as large number of terminals, dispersed distribution, wide coverage, and short communication distance. With the continued urban sprawl, some changes will need to be made to the architectural structure, the indefinite topology structure. It is, therefore, unrealistic to build a new private communication network for monitoring conditions, considering construction costs and economic benefits.

Generally, the automation system monitors and controls devices, including those installed in the switching station and distribution room. Each subsystem of the DMS is designed with different emphases. For instance, the circuit breaker tends to transmit tele-signal and tele-control signals due to its frequent operation. To satisfy these requirements, research institutions and manufacturing organizations at home and abroad have carried out much research on communication via distribution networks.

Nowadays, no single communication mode can satisfy all the requirements of automation, so it is necessary for a proper communication system to mix multiple communication channels. Under these circumstances, each channel should be applied on different occasions depending on the existing data transmission dielectric and communication mode by comparison of technology and economy. The aim here is to increase the data transmission rate and stability, and reduce construction costs as well. Different from a transmission network, a distribution network has large numbers, a wide variety of nodes, and millions of addresses, but it has a relatively short channel distance and low efficiency. This means that fiber-optical communication, radio communication, and carrier communication for distribution lines are all applied in network communications. The wide variety of functions, varying from auto-reading to load control, feeder opening/closing to substation automation, determines that communication should be reliable, economical, and have a large address space and two-way communication for convenience of operation. These characteristics are true of distribution equipment for monitoring conditions. Combined with years of working experience, distribution communication should rely on an automated communication network to build a communication system for monitoring equipment conditions.

For a distribution network device, after condition-based monitoring information access to the automation communication network, the original system will be affected in the aspects of communication bandwidth and security.

1) *Communication bandwidth.* The monitored data includes real-time state and recorded data. The monitoring information is less than the automation information in the aspects of real time and number of points of information. For instance, a transformer DGA transmits information, which is smaller than 10B every time, and once a day in normal conditions and once an hour in case of faults. The recorded message, though large, can be transmitted by subcontracting with low priority in the form of documents to reduce the impact on distribution network automation communication. So, the information transmission monitored by distribution equipment has little influence on the network bandwidth.

2) *Information security.* The condition monitored information falls within Zone III by security. While the automation monitoring information belongs to Zone I, they should not be transmitted in the same network. The condition monitoring information should be transmitted via a communication channel and terminal layer as Zone 1, and then restored to the corresponding safety zone via the information security unit near the master station. In so doing, security problems in the interconnection transmission are avoided.

Briefly, it is feasible to implement monitoring conditions with the support of an automation system communication network. It is worthwhile noting that this does not influence its real-timeness or security.

6.2.3 Primary Station for Distribution Equipment Condition-Based Maintenance and Integration of DMS

To make sure the DMS system is reliable, flexible, and transferable, the DMS system is designed to be open, distributed, and have consistent commercial database and real-time data, advanced graphic display and multimedia technologies, and extendable advanced applications. Among these, many are also needed by the distribution equipment conditions monitoring and condition-based maintenance systems. These systems rely mainly on the DMS master system so as to avoid overlapping investment and improve system integration.

Distribution equipment condition monitoring, condition-based maintenance system, and DMS are integrated to build up a unified management system. The systems in a layered and distributed design are standardized, real time, extendable, and maintainable.

1) *Multi-layered structure.* The host station system should be in a component-based design, and divide run units (components) into different layers by function. It is typically divided into an interface layer (interaction with user), an application service

layer (also business logic layer), and a data service layer (database centered, to save inside data).

2) *Open.* Openness means that applications are open in respect of portability, exchange-ability, adaptation to future development, and protection of original investment. It also includes computer, software, database, user-interacted interface standards, and various other open international standards. For instance, the operating system complies with IEEE, POSIX, and OSI; the database interface complies with SQL; the computer communication protocol is TCP/IP.

3) *Standardized design.* The standardized design is specific to all interfaces on the component of the master station system to support interactions among components. The operation standard separates the interconnected components and removes the barriers of language and physical distribution, laying foundations for a management master system.

4) *Capacity margin.* As for the capacity margin, the status quo and vision of the distribution network should be considered with adequate margins for scale, capacity, processing speed, and CPU load to make sure that the host system maintains its performance despite changes in capacity and structure. In addition, as functions and scale expand, seamless expansion is made available to the master system.

5) *Real-timeness.* This means that the host system should satisfy the requirement of real-timeness in supervisory control and management. The integrated functions of monitoring and maintenance should not affect the maintenance performance of distribution network automation.

6) *Expandability.* Good maintenance means that the system boasts high availability. The master system integrates all the various supervisory controls and meets the requirements of management, since it is highly integrated and complex. The maintenance of one local subsystem must not affect the normal operation of the remaining subsystems.

7) *Advanced graphic display and multimedia technology systems.* The host station should use rich and customized graphic display technology and various alarms, and offer working interfaces to achieve multi-dimensional directions, clearly visible regular monitoring functions, and intelligent warning forecasts, convenient controls, exception hints, and troubleshooting.

6.2.4 Geological Information-Based Distribution Network Condition Monitoring and Maintenance

The geographical information system (GIS) is a computer technology system that provides geographical information about space and dynamics, and serves for geography studies and decision-making by way of geographical model analysis based on a geographical space database. Currently, in the DMS master system, AM/FM/GIS are tightly integrated with the distribution SCADA system and have become the core architecture of the system, together forming a standard configuration.

Owing to the large number of distribution devices and dispersed distribution, the operation of distribution equipment has much to do with geographic position and the management work is very challenging. Therefore, a complete distribution equipment condition-based monitoring and maintenance system inevitably requires GIS to provide information regarding space that is monitored, analyzed, and processed by devices and display the conditions of the distribution equipment and architecture, and the dynamic operation conditions of the distribution network, in a collaborative way.

A major function for GIS is to analyze and search the space data and map it to attribute data for application of the database technique. GIS will keep the equipment condition-based monitoring and analysis updated when applied to condition monitoring and maintenance, and the space information, together with real-time information about operation, will effectively improve the power distribution and level of urgent repair. Preparing a fault diagnosis and trouble-shooting pre-plan, and improving the dispatching or maintenance persons' abilities to monitor the system and handle faults, so as to minimize losses, improves the reliability of the power system and further raises the level and practical utility of the distribution network management system and distribution equipment condition-based monitoring.

6.2.4.1 Integration Mode

The most commonly seen ways to integrate SCADA/AM/FM/GIS with the distribution equipment condition-based monitoring system are loose integration and tight integration.

6.2.4.1.1 *Loose Integration*

GIS basic software is independent of the distribution equipment condition-based monitoring system, but various forms of data-exchange channels are established between them, such as data files, database, and dynamic data exchange to achieve communication between the GIS platform and the condition-based monitoring system.

The distribution equipment condition-based monitoring system requires a GIS system to frequently exchange mass data with the real-time base, historic base, and some high applications, so the system performance will be reduced in such a mode. For a system with a commercial GIS platform, the mode of saving space data and attribute data on the GIS platform is different from that of data for general professional applications, so it is almost impossible to use one uniform mode of storage to organize the entire system application.

DMS has expanded into many professional application modules, such as distribution automation (DA) – distribution network analysis previously – which provides experience for the GIS platform to communicate and exchange data with such application modules. The practice of monitoring distribution equipment conditions could learn from the previous practices of application module access. In so doing, we can keep the achievement in applications of the distribution network and reduce the costs of system maintenance/upgrade as much as possible.

6.2.4.1.2 *Tight Integration*

With use of the powerful secondary development tool provided by GIS software, we have an efficient module for distribution equipment condition-based monitoring and analysis. In the distribution devices condition monitoring system, many modules need the support of the GIS system, so it is established in all application modules for condition-based monitoring and maintenance on the GIS platform. In so doing, monitoring, maintenance of distribution equipment conditions, and emergency integration of the GIS system are all possible, reducing data redundancy as well as development/investment in the private platform of the condition-based monitoring system.

Since foreign GIS platforms are not specifically designed for power systems, and the commercial GIS platform ignores the characteristics of electric GIS with the goal of satisfying all varieties of application, there are many functions which will not be used in electric GIS. Neither can it meet the requirements of a distribution automation system in respect of performance and efficiency. However, the distribution device condition-based monitoring and maintenance system requires little in real time, so it is easier to integrate the system.

6.2.4.1.3 *Comparison of the Two Methods*

In the mode of loosely coupled integration, the condition-based monitoring system of the distribution network is loosely coupled with the AM/FM/GIS systems. As an independent system, the condition-based monitoring system not only satisfies the requirements of regular distribution equipment but expands its monitoring objects from relatively concentrated devices into the feeder, switch station, and other dispersed devices. The condition-based monitoring system is integrated with the AM/FM/GIS or other systems by exchanging real-time data.

In the tightly integrated mode, the distribution device condition monitoring system and AM/FM/GIS systems are deemed to be a whole. The condition monitoring subsystem has functions of data acquisition and supervisory control – i.e., no maintenance and operation screens, no network modeling. What it does, as a background component, is transmit real-time data to the AM/FM/GIS system and receive orders from the AM/FM/GIS system.

Each integration mode has its advantages and disadvantages. In the tight integration mode, since data maintenance and management are performed in a centralized way, both AM/FM/GIS and condition monitoring or other advanced applications work under a uniform interface, the AM/FM/GIS system is required; otherwise, test and management functions will not be realized.

For technical and investment reasons, most of China's existing distribution management systems, distribution condition monitoring, and AM/FM/GIS systems are loosely integrated. Regardless, as the demands for integration increase, a more closely integrated mode is the desired trend in future.

6.2.4.2 Information Interaction

The information interaction between the distribution equipment condition-based monitoring system and the GIS varies from one integration mode to another.

In the loosely integrated mode, data are one-way exchanged. The distribution device condition monitoring system is able to finish condition acquisition and estimation, while the AM/FM/GIS system manages the distribution network. The system provides real-time and historic data to the AM/FM/GIS. In turn, the AM/FM/GIS does not have to provide the condition monitoring system with data.

In the closely integrated mode, data are two-way exchanged. The distribution device condition monitoring system provides real-time and historic data to the AM/FM/GIS and receives orders from the AM/FM/GIS.

In actual applications, the following data should be exchanged.

1) Information from the FIS, including geographic background figures, off-line graphic data, and distribution parameters.

2) Information that the GIS obtains from the distribution condition monitoring system and displays in a graph, including:
   - real-time operating information – dissolved gas in oil, partial discharge, iron-core grounding current, contaminated current, temperature and humidity;
   - real-time running information on the potential transformer (TV), current transformer (TA), coupling capacity (OY), capacitive voltage transformer, leakage current, nominal capacitance, equivalent capacitance, and dielectric loss;
   - real-time running information on the capacitive device – primary leakage current, resistance current, fundamental current, power loss, and busbar voltage;
   - real-time running information on the MOA – primary leakage current, resistance current, fundamental current, power loss, and busbar voltage;
   - state of insulating devices – normal, insulation aging degree, and maintenance state.

   After the information exchange, using the measures provided on the GIS platform (such as view and table, as well as dynamic coloring, and refreshing technologies, with the coordination of multimedia sound and light signals), condition information can be displayed completely, accurately, and visually.

3) Fundamental functions of the integrated system, including device condition monitoring and GIS, of layer management and real-time monitoring. The layer manager manages the layers of the system and subsystem, such as definitions of names, symbols, rendering, implicit/explicit, modular sequence control, scale density control, and label special control. A geographic map is usually displayed as the background layer, building layer, river layer, and other ground-feature layers. Distribution devices, if layer displayed, include LV power grid, MV power grid, tower pole, switch, and transformer layers. The detailed division of layers depends on actual needs. The sequence of layers can be adjusted by dragging. Map functions include navigation, zooming, and roaming. The navigation identifier identifies the coverage of the current map and positions the location on the navigation map or quick positions by selecting a range.

Condition data is displayed in real time via a real-time module; the running management module and subsystem ports and combines the data with geographic information in the proper way.

The GIS serves as an analyst of the data detected by the distribution devices. The GIS application features query and analysis by regular conditions or by attribute, conditions, or locations, such as the transformer information of a region or the load data of an urban area. The GIS system also acts as an aided supporter of equipment maintenance, like offering the best rush-repair path.

In addition, the GIS user management and security log provide password protection for all kinds of operation, and the system provides a user management interface. The administrator has the authority to authorize, add, delete, and modify users to ensure system safety and data correctness. Important operations should be noted for research purposes in case of fault. What's more, the system also provides distance measurement, area computation, and interface customization.

6.2.5 Distribution Equipment Assessment and Condition Maintenance

Condition-based monitoring is designed to support the management of the distribution network and reduce the workload of equipment maintenance. So, it is necessary to construct an on-line monitoring system in DMS and build a condition-based evaluation maintenance and decision-making system. Its functions include:

1) getting and processing characteristic parameters that may reflect healthy conditions of equipment, such as basis, real-time, and historic data for devices;
2) evaluating the current condition of devices and predicting future development trends;
3) giving warning information and making fault model and cause analysis, once there are exceptions to devices;
4) supporting the implementation of operation and maintenance, with the model analysis of strategy for optimizing maintenance all round.

Based on the automation system and GIS, the condition-based monitoring and maintenance system is able to improve both power supply enterprises and distribution networks in many ways:

1) it makes a timely estimation of conditions and risks so as to make a pre-plan for DMS and prevent trouble before it happens;
2) it provides aided decision-making for post-failure fault diagnosis and shortens the time needed to remove faults;
3) it is able to make an appropriate estimation of equipment conditions and improve the condition of the distribution network.

As a result, in order to improve the security of a distribution network or optimize the operation of a supply network, the condition-based monitoring and maintenance system should be integrated with functions such as condition-based monitoring, estimation, risk assessment, fault diagnosis, and maintenance plan, based on the device management information system.

6.2.5.1 Information Support

Device conditions should be estimated based on complete information, including routine and regular inspection, diagnostic tests, on-line monitoring, live detection, family defects, and poor working conditions, especially their seriousness, magnitude, and development trends. To foster effective management, a perfectly functional distribution network management information system should be built.

Currently, China's advanced distribution network management system involves power-supply enterprise operation management, equipment management, and customer-service systems. Based on the automation real-time environment, GIS, and integrated database system, the system builds several independent application function subsystems, including distribution work management (DWM), distribution automation (DA), automatic mapping (AM), trouble complaint management (TCM), facility management (FM), load management (LM), and distribution analysis system (DAS), to make automatic management possible and improve the operation and reliability of power supply.

Currently, the distribution facility management subsystem is widely used in the electric power industry, so that almost all enterprises with information construction are equipped with it. Other representative functions include ledger management (including add, delete, modify, inquire, and browse), user information and authority management (including user register, assign authority, modify user information, delete user), facility defect management, equipment maintenance warning, query, analysis and statistics of operating information, routine test management (including routine test log, inquire, statistics, print table), report display, comprehensive inquiry, and print.

The present device management subsystem for distribution networks has played a certain reference role in improvements at the device operation management level. However, it still has some problems if using this subsystem to optimize the trouble-shooting system. This is because the device management subsystem mainly puts emphasis on asset technology management, including device ledger and off-line monitoring data management and maintenance activity process management. The main maintenance measures refer to traditional planned periodic maintenance and posterior maintenance, which excludes the real-time monitoring and assessment of device health as well as the guidance of maintenance activities. As an independent information system, the existing on-line monitoring system is usually used to monitor some single device index. The monitoring and diagnosis items are incomprehensive, and the diagnosis method is also relatively simple. So, it cannot effectively be extended. All of this results in the information related to device maintenance being dispersed in different subsystems; the application system environment being different; a uniform data format and data transmission system lacking; and the integration level of the system being poor, which further causes the system to be impractical for a maintenance system. Furthermore, although some information systems developed at home and abroad can be used to realize uniform maintenance management, their strategy implementation lacks consistency, the systems are not strongly developed, and their extension and upgrade are both difficult. Therefore, such systems cannot adapt well to changing business processes.

Comprehensive and systematic data serve as the basis for device condition estimation and condition-based maintenance. The condition analysis, diagnosis, and trouble forecasting need to be incorporated with on-line monitoring data, off-line test data, device technology ledger, maintenance log, and exceptions records. To optimize maintenance, it is necessary to make a proper schedule for maintenance items and time (i.e., maintenance decisions and schedule). This requires condition, diagnostic information, and statistical data from the same type of facility, maintenance supplies and materials, personnel, basic parameters for the power grid, future load data, and mode of operation, with all these provided as the basis for decision-making. At present, the application systems are from different manufacturers and have been developed independently. What's more, they use different system platforms and data structures, so these subsystems are independent islands of automation, being characteristic of distribution and heterogeneity. Therefore, information sharing and coordination among systems cannot be achieved.

In order to establish the information foundation for condition assessment and maintenance, domestic research institutes and enterprises carry out a great deal of work. For example, Condition Grid Henan Electric Power Company establishes codes related to information integration and information flow management for the device maintenance management system, device health information management system, device operation management system, and turbine on-line and off-line diagnosis system of Yaomeng Power Plant and integrates the above systems at the data layer. However, the openness and expandability of the system is not good, and its following protocols are also exclusive.

It is necessary to collect static and dynamic data for device condition assessment and maintenance of the distribution network. The static data includes the natural characteristics of the device, such as device nameplate, manufacturer, factory experimental data, and so on. Then, the dynamic data involves the real-time operation condition of the device, for example, real-time monitoring data, chromatographic sampling data, periodic test data, and failure of device. Such information is distributed in application systems developed by different manufacturers. Q/GDW 382-2009 *Technical Guidance for Distribution Automation* defines the distribution network information exchange mechanism (i.e., the mechanism based on information transmission). This mechanism must be used to realize the exchange of real-time, quasi-real-time, and non-real-time information, support the business transfer and function integration among multiple systems, and make the automation system of the distribution network share information with other related application systems. Such information exchange should follow the common information model (CIM) in IEC 61968, especially the device information management subsystem for a distribution network on the basic information platform. The data acquisition of advanced applications (e.g., condition assessment) will access this basic maintenance data through the standard access interface in IEC 61968. These standard access interfaces should be able to access the distribution network model, real-time data, event and alarm data, and historical operation data.

6.2.5.2 Distribution Device Condition Assessment

Device condition assessment means using the corresponding off-line and on-line data to obtain the operational performance of a distribution network device and calculate its service life expenditure. Based on the realization of condition monitoring and device information management, this function is used to evaluate the damage and expected service life of operational devices. The maintenance time of a distribution network device is closely related to its service life. If a model can be established for the condition of distribution network devices and real-time assessment can be achieved, the optimal dispatching of distribution networks and optimal maintenance of distribution network devices can be fully realized. Therefore, an accurate assessment of device condition is the basis for improved device operation mode, safety of distribution network, and optimization of operation and maintenance work of power-supply enterprises.

The condition assessment of a distribution network device is based on the DMS platform. It is necessary to establish various device service life models based on historical data and family information, calculate the limit of electrical endurance of each kind of device through the on-line monitoring information of the distribution network device (with the help of the distribution network terminal), and forecast the service life of the device (including electrical life and mechanical life).

Device condition is assessed according to the ledger, preventive test data, operation data, historical fault record, tour inspection record, on-line monitoring data, and other related data of the distribution network device. The condition assessment refers mainly to the following processes: first, effectively collecting and organizing device operation information; then, judging the current operational condition of a device according to the assessment standard and system of device working conditions to see whether the device is operating normally; finally, guiding the related production operation persons to make a device maintenance plan based on previous judgment conclusions.

The current operational condition of the device is the most direct basis for condition assessment. However, due to environmental changes, device differences, monitoring means, operational management, and other related reasons, device condition based only on current real-time monitoring results cannot usually be judged accurately. The assessment should be based on tour inspection, routine tests, diagnostic tests, on-line monitoring, live monitoring, family defects, poor working conditions, and other related information, including degree of phenomena, magnitude, and development tendency. The operational condition and health level of devices can be known accurately only through the daily monitoring and inspection of operational conditions, comparison of partial on-line monitoring devices with similar devices, as well as continuous and normative follow-up and management of all off-line and on-line analysis results (through which we can make an effective judgment, thus laying the foundation for the next stage of maintenance work).

During the evaluation of equipment conditions, if the tested value or detection value has a tendency to exceed limiting values or almost reach alert values, then the evaluation should keep track of the on-going equipment; power failure equipment should not be put into service if it is suspected of major defects. For values set with alarm limits, if there is an evident tendency for the test value or monitoring value to be close to or exceed the alarm value, then the running equipment should be stopped for a black-out test as soon as possible. The power failure equipment must not be put into service before the hidden dangers are completely removed.

Besides, in similar operational and inspection conditions, the same state parameter of devices of the same family should not have any visible differences. Otherwise, significant difference analysis should be carried out.

The condition assessment is carried out according to guide rules and standards related to condition characteristics and assessment of power transmission and distribution network devices, which analyze and evaluate all indexes and data related to device health and finally obtain the overall condition level of the devices. Since the condition assessment of distribution network devices is a multi-attributive problem, it is necessary to classify the condition of devices reasonably comprehensively considering the monitoring data, working environment, and operation maintenance record, establish a comprehensive assessment index system, and gradually correct this system based on experience gathering, thus improving the accuracy of assessment step by step.

Take a transformer, for example. Various conditions of transformer can be obtained by comprehensive assessment of all indices. When the indices exceed the limit, comprehensive estimation should be dropped and the condition of the transformer should be judged in the following ways. For instance, when the absolute value of one or more unimportant indexes exceeds the limit, if it shows no obvious trend of the condition worsening, we could say the transformer is under an alert condition. In another case, where one or more indexes (such as winding dielectric, DC resistance, or content of acetylene) exceeds the limit, the transformer is in danger, which suggests that there is trouble in the transformer. At the time, a maintenance strategy is determined depending on the trend of degradation. When there is a marked trend of degradation, the transformer should be terminated for maintenance immediately. As a result, if we make a reasonable selection of condition-based monitoring information and operating maintenance records, build an operation state estimation model by module, and compound rules with the improved evidence, we will have access to results on condition estimation.

The data input of the condition assessment function includes the following.

1) *Criteria for condition evaluation.* According to the *Guideline for Condition-Based Maintenance of Oil-Immersed Transformer and Reactor,* we should determine the condition variables in the first place and express the evaluation criterion for each variable as a form of database. Among them, the expression for evaluation criteria in the form of a database is one of input data.

2) *State assessment data (nested).* This refers to the processed values of the state variable that are obtained from the on-line monitoring system, maintenance tests, daily maintenance, and other data sources.

3) *State variables data.* This refers to the values of state variables obtained from basic data (ledger, factory default value) or via another channel that may reflect the initial state and be used for evaluation.

The condition assessment function includes the following.

1) *Evaluation criterion for maintenance devices.* This can modify the evaluation standard module for all devices that are built up in accordance with state characteristics of transmission and transformation devices and relevant guidelines.

2) *Evaluate the healthy condition of parts.* According to the condition assessment standard, build an algorithm model for healthy condition evaluation. With comparative results of horizontal equipment of the same type and longitudinal historic data, grade the state variables one by one that have an impact on the healthy conditions of all parts and evaluate the healthy conditions of all parts of devices in a quantitative way.

3) *Assess overall healthy condition.* Based on the result of condition assessment, combined with the impact of parts on the general function, we obtain the overall healthy condition level with a proper algorithm.

4) *Assessment information query.* This can inquire about the condition assessment results of devices and parts and provide details about the assessment process and all condition variables.

The condition assessment output refers to the results of condition assessment, including healthy level or score, basis, and explanation.

6.2.5.3 Device Risk Assessment

Risk refers to an index used to evaluate the possibility and seriousness of damage, while risk assessment refers to a process by which the risks that devices face and may lead to are determined. The risk assessment has its origin in reliability theory, and is based on an analysis of probability. Risks are unavoidable. When devices are deteriorating, or there are signs of deterioration, the risks are increased if they continue to run. At the time, immediate measures should be taken to prevent loss. So, an optimization policy on condition-based maintenance and the maintenance plan should be made on the basis of risk assessment, so it will be targeted and more reliable to ensure reliability and save costs.

In the past, to ensure the availability of the electric power system, China's electric power enterprises have improved the qualities of individual devices and strengthened maintenance with a view to ensuring the availability of both subsystems and the power system as a whole, with little consideration for the analysis of return on investment.

Now, due to the large numbers of power grids and strict cost controls, power-supply enterprises should introduce risk analysis technology to device management and carry out a strategy of asset management based on risk evaluation. When the reliability of a power system is guaranteed, enterprises can make efforts to reduce the costs of operation and maintenance, and increase the return on investment:

In accordance with IEEE 100-1992, widely applied to the management of engineering equipment, risk is defined as a concept and measurement of seriousness of unwanted consequences, expressed as a product of probability and consequence. It is expressed as

$$R = PC \qquad (6.48)$$

where R is risk, P is the probability of an event, and C is the consequence of the event (or loss).

To evaluate risks, first determine the definition and model of uncertainty and adverse consequences of the risk events on the researched object. The variables are defined as either qualitative or quantitative.

The uncertainty of risk events can be *quantitatively* described as rare, unlikely, possible, very likely, affirmed. The analysis of quantities is often described as a probability in terms of statistics, ranging from 0 to 100%.

The adverse consequences of a risk event are divided into the loss of many factors by the researched object – such as direct asset loss, personal security, loss of health, environmental pollution, and loss of reputation. With the losses and function relation, a total loss will come out. The losses can be classified into very small loss, small loss, moderate loss, big loss, and catastrophic loss. The losses can be calculated in currency terms during the quantitative analysis, and the losses of all factors are finally converted into economic losses.

At present, many mathematical methods are available to describe the relationship between the uncertainty of various risk events of an object and unwanted results. In this manner, risk assessment is usually calculated by a risk coordinate method, Monte Carlo method, failure mode and impact analysis, or event tree analysis. The selection of technical methods is made depending on the condition of the estimated system and the risk control goal.

Nowadays, the risk assessment technology has made preliminary progress in European countries and been applied to the asset management of power grid enterprises. Between 2004 and 2005, Shandong Electric Power Corporation (SEPCO) and Austrian TransGrid developed wide cooperation in the study of a risk assessment system for transmission and transformation devices, and completed risk assessment and ranking for 110 kV and above transformers of SEPCO.

In September 2007, SGCC issued *Guideline for Risk Assessment of Transmission and Transformation Devices* (request for comments) (2008) No. 32 to provincial power companies, followed by a trial standard in March 2008. This document describes the model, procedure, and method of risk assessment for transmission and transformation devices. At present, research is underway by all power companies.

According to the *Guideline for Risk Assessment of Transmission and Transformation Devices* (trial) (2008) No. 32 released by the State Grid, assets of devices, losses of assets, and probability of faults are taken into consideration, building up a risk assessment module and the risk caused by faults in transmission and transformation devices in a quantitative way.

1) *Model.*

$$R = AFP \tag{6.49}$$

where A is an asset and F is asset loss (device damage, personal safety, reliability of power supply, social influences).

2) *Assets.* Asset A is the weighted value of device asset A1, user level A2, and device position A3, varying from 1 to 10.

3) *Losses of assets.* To calculate F, we should first introduce the index of factor (IOF) and probability of failure (POF). IOF is defined according to the criteria for accident classification, for instance, an extra-serious grid accident is expressed as Level 9 in terms of IOF; POF refers to the probability of losses in terms of costs, environment, personal, and power grid securities that come with the fault occurring in some types of device and can be calculated according to statistics on historical fault data. The loss degree of assets F is the weighted value of loss degrees of costs, environment, personal, and power grid securities. For some type of transmission and transformation devices, F will arrive at a value according to historical fault data.

4) *Probability of fault.* It is uncertain whether faults will occur. The probability of fault varies from one device to another. Even for devices of the same type, not all probabilities of fault are the same due to differences in manufacturing process and operating environment. For this reason, we can get the probability value according to the formulation by the collection and analysis of historical data. In accordance with the *Guide for Risk Assessment of Transmission and Transformation Devices*, the mean failure rate can be calculated according to equation (6.49). Then, with the statistical analysis of conditions and mean failure rates of devices in power grids under the supervision of provincial power companies, determine the proportional coefficient K and the coefficient of curvature n.

Inspired by the *Guideline for Risk Assessment of Transmission and Transformation Devices* (trial) released by the SGCC, the results of risk assessment can be expressed using a quantitative method. With value-at-risk as one index, the risk assessment can come to a conclusion by taking into consideration the interaction of assets, losses of assets, and probability of faults.

The risk assessment module will calculate losses of assets and probability of occurrence once devices are subject to failure threats by sensing internal defects and external dangers. Through the module, we obtain the risk level.

The risk assessment function inputs data, such as the result of state evaluation (score), cases of equipment failure (device failure, losses, and probability), and equipment information (including equipment ledger, power grid structure, and users).

Goals of risk assessment:

1) the asset class of electrical equipment is determined depending on the value of the device itself, the importance of users, and the importance of devices in the power grid;

2) defect identification and threat analysis to identify potential internal defects and external threats and calculate probabilities of occurrence in the method of statistical analysis;

3) losses caused by threats calculated through associated devices and defects, or threat factors in the aspects of safety, reliability, costs, and social influences;

4) value-at-risk or score calculated by taking all this into consideration, including asset class, loss of assets, and probability of occurrence;
5) risk assessment inquiry to support the inquiry of risk assessment and provide detailed information about the process of evaluation and indexes;
6) risk assessment output to provide conclusions of risk assessment, including value-at-risk (score), basis, and explanation.

6.2.5.4 Fault Diagnosis

The purpose of fault diagnosis is to find the reasons and locations of failures by mathematical calculation and logical deduction from the states of related equipment, expertise, and experience on the basis of condition-based monitoring and assessment. Fault diagnosis is at the heart of the optimization and maintenance of devices.

Fault diagnosis uses a comprehensive technology, involving subjects such as modern control theory, signal processing and model recognition, decision science, information science, artificial intelligence, technology of electronics, mathematics with statistics, and fuzzy logic. As a modern engineering discipline, fault diagnosis attaches importance to not only theoretical research but also applications. In recent years, great efforts have been made by academicians at home and abroad to improve fault diagnosis in electrical equipment.

According to different standards, there are several classification approaches and each has its own advantages. Generally speaking, there are basically two approaches to classification: mathematical model-based and artificial intelligence-based approaches. The former has indeed made significant progress, such as Kalman filtering, observer, parameter estimation, and parity space. Since a very accurate model is a must, limitations exist in the application of the technology. The latter is becoming more and more important, since no accurate model is required. The intelligence is reflected in the ability to diagnostically deduce (combined with the knowledge of experts in the field and artificial intelligence technology) in the process of diagnosis, to identify and predict the state of objects under the given circumstances, and to rapidly analyze and diagnose multi-fault, multiple-process, and abrupt faults with best practice in multiple fields. An intelligent fault diagnosis system is characterized by:

1) artificial intelligence, used to simulate man's logical thinking process, to solve complex problems that require logical deduction;
2) search for and capitalization of experts' knowledge and experience according to actual needs for the sake of diagnosis;
3) self-learning and self-improving, with the exchange of information under the same conditions, to acquire new knowledge from the changed conditions, organize, modify, and maintain knowledge in the database.

Fault diagnostic methods generally include expert systems, artificial neural networks, fuzzy set theory, wavelet analysis, genetic algorithms, analysis theory, rough set approach, and fault tree models. Each technology has its own advantages and disadvantages. They may improve system performance in some aspects and to some degree, but there are problems that are difficult to solve in the system design and implementation. Systems as complex as electric power systems are not likely to be diagnosed with reasoning control strategy alone. With the analysis of characteristics of subsystems, one or a combination of more technologies is adopted.

A fault diagnosis module is used to find the possible reasons for faults and their locations on devices in poor condition according to state assessment, and provide a basis for fault removal or recovery. The data input includes state variables, complementary state variables, results of state assessment, and data nested/diagnosis rules. The module is capable of the following.

1) *Maintaining fault diagnosis rules.* It creates a rule base for fault diagnosis by fault type and complexity, accumulates knowledge, and expresses the diagnostic rule in the form of a database.

2) *Developing a condition-based diagnosis algorithm.* A modifiable, extendable, plug-in algorithm is developed that satisfies the current business in accordance with the requirements of condition-based diagnosis. Based on the comparative results of state variants on horizontal (equipment of the same type) and longitudinal (historical data) coordinates, an efficient diagnosis algorithm is implemented, such as a three-ratio method, graphic method, electrical research method, neural network algorithm, or decision tree method. The diagnostic algorithm base should include the necessary expert knowledge base.

3) *Condition-based diagnosis.* Condition diagnosis is started in an appropriate way (manual or triggering mode). If there is a fault, its development trend, possible causes/locations, and bases/explanations are found.

4) *Querying diagnostic results.* Through this function, users can inquire about diagnostic results and the basis on which those results depend, tracing back through the process of diagnosis.

5) *Fault diagnostic output.* This refers to the results of condition-based diagnosis, including (if there is a fault) its fault development trend, possible reasons, location, basis, and explanation.

6.2.5.5 Condition Improvement and Maintenance

China's electric power industry has long implemented a post-maintenance and preventive scheduled maintenance system. The scheduled maintenance system has taken an important and positive role in the prevention-oriented maintenance system, which is classified into overhaul, minor repair, and temporary repair. The advantages lie in ensuring a stable power supply and that scheduled labor, supplies, and funds are well organized. In the maintenance system, the maintenance items, duration, and period are all determined by management, as the case may be. Nonetheless, there are defects in the maintenance system of China's electric power industry. Some of these are listed below.

1) *Deficiency in maintenance.* This is where faults occur before the maintenance date for some reason, and operators fail to immediately find out the trouble in a preventive test, due to the constraints of the maintenance schedule. The maintenance costs will be increased and unnecessary accidents may occur should the faults continue and worsen.

2) *Excessive and blind maintenance.* Here, scheduled maintenance follows the rule that timely maintenance is required, and maintenance is not finished until all problems are fixed. No matter whether it is faulty or not, a device is expected to go through periodic maintenance. This is an evidently blind practice. Despite the requirements and principles, there are varied problems in fact. For devices in good order, an overhaul after breakdown may affect the performance or shorten the service life. It may cause unnecessary outage costs, waste of human, material, and financial resources, and even failures.

3) *Increasingly conspicuous defects.* With the growing demands for reliability, power companies have increased their investment in maintenance and defects hence become increasingly conspicuous. It is imperative to implement a more advanced and scientific management and maintenance system, which will benefit both the power company and the community.

In traditional maintenance systems, the arrangement of the schedule is at the center of the system. According to the maintenance schedule, preventive maintenance or post-maintenance is carried out to fully prepare for maintenance, electric network transformation, and defect treatment. At the same time, it makes sure that comprehensive maintenance is performed in an orderly manner, to prevent further power outage. Maintenance schedules are divided into yearly, quarterly, and weekly plans by the planned time.

The maintenance schedule still plays an important role, although a condition-based maintenance system has been introduced. Owing to the particularity of the electric power system, some devices (especially key facilities) make a difference to the mode of operation and power supply–demand relationship. Demand for maintenance of other devices, especially devices that carry out periodic maintenance, should be taken into consideration before working out a proper maintenance schedule to ensure continuous supply of power. In this sense, condition-based maintenance has its own plan or schedule.

Maintenance modes include condition-based maintenance, periodic maintenance, and post-maintenance. These methods are implemented based on condition-based monitoring, condition-based assessment technologies, and a clearance information system. Optimization of the maintenance schedule is at the center of improving device maintenance. It is generally adopted to vigorously promote condition-based maintenance, combined with post-maintenance, so as to make systems more reliable and reduce costs.

There are many uncertainties in drawing up a device maintenance schedule, which is designed to achieve multiple goals with a lot of restrictions. Operators find that the difficulties of optimization lie in multiple dimensions, mixed integers, and nonlinearity. Considering the important roles of device maintenance programs and solving problems with an optimized algorithm based on theory and practice, scholars at home and abroad have put much emphasis on research into the mathematical model and optimization algorithm. These models are designed to achieve realizable and economic goals in respect of knowledge; the models are divided into certain and uncertain models with respect to analysis. In terms of solving problems, the approaches are divided into mathematical and artificial. As a maintenance program to improve integer and high latitude, a great deal of coverage is available on traditional mathematical approaches, each with their own advantages and adaptations. As the system expands, more devices are scheduled to go through maintenance, which brings about a problem – the curse of dimensionality. An intelligent algorithm can solve the above problems to a certain extent, and construct an efficient and smart algorithm to optimize the maintenance program, promoting improvement both globally and locally. This is still a subject badly in need of research and study.

In practice, it is worthwhile promoting the reliability-centered maintenance program. This is because a distribution system becomes less reliable once the devices are stopped for maintenance, while the maintenance runs are far less sensitive to duration than generator unit.

Reliability-centered maintenance (RCM) serves as a system engineering approach to determine preventive measures and improve maintenance systems worldwide, and is the preferred method used by national military and industry sectors to develop an outline for military equipment and preventive maintenance. According to the definitions of functions, faults, causes, and effects in RCM, the system revises faults through its fault model and effect analysis (FMEA), and lists all functions, fault models, and effects analysis, and estimates the consequence of faults by classification. Ultimately, decisions on whether to take preventive measures or repair after a fault occurs should be made.

The module of the maintenance plan with reliability orientation is made depending on the fault rate and consequences of the fault, with the goal of achieving reliability; the schedule of device maintenance is made in an optimized manner. The rate of fault has to be given top priority. Traditionally, it was common practice to assess the rate of fault of a reliable model, as being representative of devices of the same type. This is too simple to include the actual conditions of unreliable devices. At present, foreign scholars have developed a method to calculate the rate of fault according to a functional formula, which takes service period and history of maintenance as functions, and creates a modeling method for calculating the rate of fault based on data about the condition of the equipment. This kind of calculation is carried out by: (1) making an assessment of each index based on the data from a preventive test; (2) acquiring a comprehensive score by weighted average; and (3) converting the condition score into a rate of fault through the exponential function. In practice, a reliability-centered maintenance plan under the guidance of the failure rate generated by the state of a device embodies the goal of improved maintenance practice. In addition, in the reliability-centered maintenance plan, most research literature considers only influences that stopping devices for maintenance bring about on the system, ignoring influences on the device itself. In fact, maintenance work improves the condition of transmission and transformation devices, and reduces the possibility of fault occurrence. So, the influence of maintenance on the probability of faults is considered to be a reliability index.

Currently, we are still unable to apply the condition-based maintenance system in the State Grid, especially the distribution network. For comprehensive efficiency, it is necessary to improve the procedures for maintenance based on all kinds of existing methods of maintenance, in combination with the characteristics and needs of the distribution network, in an appropriate maintenance strategy. As for the selection of the maintenance strategy, it is determined by taking the importance, fault type, diagnosis, reliability, technology, and economy into consideration. As a rule, breakdown maintenance is possible for devices whose fault consequences are not that serious. If the conditions of devices are not available easily, or the costs of condition-based maintenance are far larger than the benefits from it, preventive maintenance is the best choice. For devices equipped with a condition-based monitor, with serious consequences of failure, condition-based maintenance is preferred.

The central idea of improving the maintenance level of the State Grid is to gradually replace periodic maintenance with condition-based maintenance, accompanied by a comprehensive plan for reliability-centered maintenance so as to save the costs of repeated breakdown maintenance and ensure the continuity of electric power system. It should be mentioned that condition-based maintenance is the core and basis for improving the maintenance of the distribution equipment. Therefore, the distribution network will put emphasis on the construction of condition-based monitoring and maintenance systems for the next stage.

7

Implementation of Self-healing Control Technology

A distribution self-healing system should not put overdue emphasis on the dispatching or automation system, since a strong grid structure is the basis and an indispensable part of self-healing. The continued supply of power is essential to a self-healing system. A strong power grid, information transmission, and self-healing control system are the most important factors comprising a distribution network self-healing system.

7.1 Principle of Implementation of Self-healing Control

7.1.1 Characteristics of Self-healing Function

The self-healing function refers to the capacity for self-prevention and self-recovery. Self-healing control has the following characteristics:

1) preventive control as a primary means of control – that is, timely discovery, diagnosis of faults, and elimination of problems;
2) capability to maintain normal operation without damage to the system and recover from faults through the self-repairing function.

A self-healing grid or self-healing control is an inevitable trend for power grid dispatching and controlling systems. Given the wide-area dynamic of an electric power system, adaption as well as coordination is vital in resolving the problems of local and global function coordination and speed.

7.1.2 Basic Principle of Self-healing Control

Self-healing control should follow the basic principles in the control logics and structure design.

1) *Distributed self-governance.* The wide-area and rapid dynamic process requires control devices to be self-governed.
2) *Wide-area regulation.* Grid security concerns the global grid, and regulation often involves (for this reason) achieving good regulation of security; efforts need to be made to coordinate the whole and the parts. Since a power grid is quick to change and develop, while the plan for a global control scheme is formulated only slowly, we must solve the problem of how to match their pace in practice for the purpose of regulation.

Self-healing Control Technology for Distribution Networks, First Edition. Xinxin Gu and Ning Jiang.
© 2017 China Electric Power Press. Published 2017 by John Wiley & Sons Singapore Pte. Ltd.

3) *Principle of industrial condition adaption.* The ability to deal with changes is the foundation of effective operation of a control scheme, and adaptive control should be based on measurement; an intelligent processing function is required for condition evaluation and control scheme.

The operation conditions of a distribution network can be defined as emergency, recovery, abnormal, alert, and safe, and there are four controls corresponding to these statuses, as shown in Figure 7.1. The purpose of control is to make sure of correct operation in case of faults, and the margin of safety is maintained in accordance with $N-1$.

1) *About prevention.* In the state of alert, possible failures are eliminated by means of inspection of the secondary system, adjustment of the reactive compensation device, and line switchover.
2) *About correction.* When the distribution network is abnormally operated, control measures should be taken to remove abnormal operation, eliminate overload and voltage off-limit, and prevent voltage instability. The state of the system turns to alert or safe condition.
3) *About recovery control.* In the state of recovery, choose a proper power-supply path to restore load supply and island parallel operation. Turn it to the state of abnormal, alarm, or safe conditions.
4) *About emergency control.* In case of emergency conditions, to maintain stable operation and continuous power supply, a series of measures – including cutting off faults, cutting machines, and system active splitting control – must be taken to transfer the system into recovery, abnormal, alarm, or safe conditions.

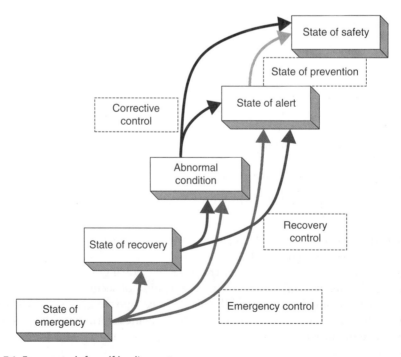

Figure 7.1 Four controls for self-healing system.

7.2   Self-healing Control Method

7.2.1   Urban Distribution Network Self-healing Control Method Based on Quantity of State

Figure 7.2 shows the process of urban distribution network self-healing control based on quantity comparison of the state. It first defines the system state functions f related to voltage, current, power, and frequency and their scope of functions $f_{ex}, f_{re}, f_{se}, f_{cr}$ in the

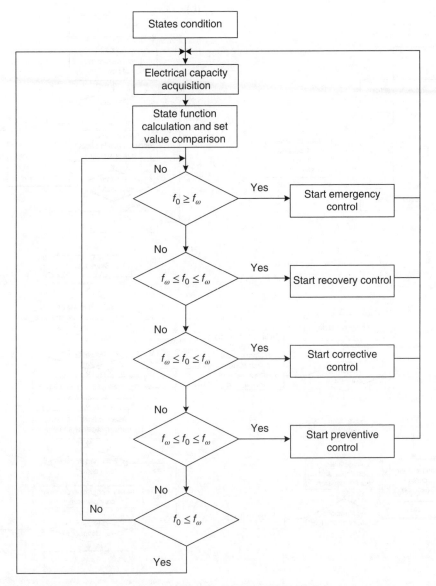

Figure 7.2 Flow chart for urban distribution network self-healing control. f_{ex}, f_{re}, f_{se}, and f_{cr} are the limit values in the state of emergency, recovery, abnormal, and alarm conditions.

state of emergency, recovery, abnormal, and alarm. It then compares the calculated state functions with the set values of the system state function to determine the operating condition and corresponding control measures, so as to enable the system to transfer to a better state.

Figure 7.3(a) shows emergency control, 7.3(b) recovery control, 7.3(c) corrective control, and 7.3(d) preventive control.

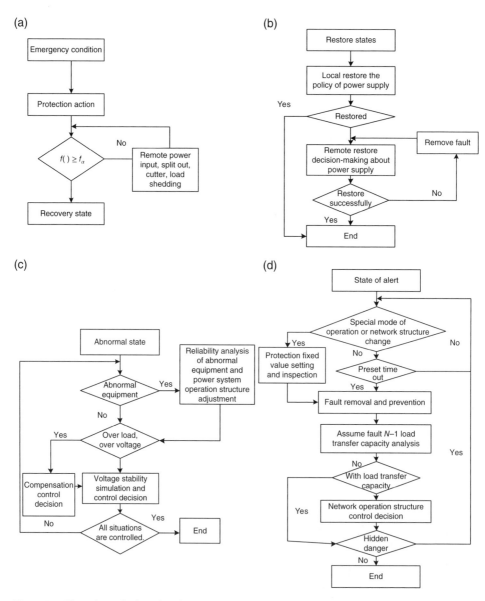

Figure 7.3 Flow chart of urban distribution network self-healing control: (a) emergency control, (b) recovery control (c), corrective control, (d) preventive control.

The self-healing control software program based on the above theory is embedded into an urban distribution network automation system and collects real-time information by virtue of automation equipment, such as a new protection and monitoring device, or on-line monitoring device. With this information, the system is able to make decisions about the self-healing strategy and upload it to the control equipment, which will coordinate generating, screening, confirming, and implementing the strategy. In so doing, it will make data acquisitions, control decisions, and actuating equipment integrations, and give the urban distribution network the ability to self-defend and self-heal. Therefore, the distribution network system has the capability to cope with extreme disasters and emergency accidents on a large-scale grid, improving the reliability of the power supply.

7.2.2 Self-healing Control Method for Distribution Network Based on Distributed Power and Micro-grid

An active power distribution network is vital to achieving a self-healing function for the distribution network. Conventional energy power can be used as emergency power supply, whereas an intermittent energy power plant lacks the capacity to serve as emergency power supply due to the instability and uncertainty of power-generating capacity and power energy. Only when the power plant is equipped with an energy-storage battery can it act as an emergency power supply or backup power source. Distributed power, relative to traditional centrally connected power, refers to 50-MV or below power plants at the load area. Generally, distributed power is either connected to the distribution network or supplies power to nearby loads independently. A micro-grid is an effective way of connecting and managing distributed power-generating equipment. Figure 7.4 shows a flow chart of distribution network self-healing control based on distributed power and micro-grid.

Figure 7.5 shows a diagram of a power island micro-grid system; the transformer is used to isolate the mains current system and convert the voltage. Sensors are designed to sense and monitor the electrical quantities of the power system, for instance, voltage, current, frequency, and harmonics. Island control switches are designed to connect or disconnect the electrical link between the island and the main AC system. A nonlinear load switch helps the controller to directly switch on/off a nonlinear load. In order to guarantee excellent performance of the distribution network in normal operation, continuous improvement of the distribution network should be made. The major advantage of the distribution network self-healing control method based on distributed power, micro-power grid, and demand-side management is that it gives full play to the important role of the micro-grid in supporting the distribution system in case of fault or being subjected to disturbances, and in continuous optimization during normal operation.

7.2.3 Distribution Network Self-healing Control Based on Coordination Control Model

Based on an integrated coordination control module, the processing signals and generating control strategies are adaptive, automatically completed without any external intervention, and coordinate with relay protection units. The automatic regulating devices and parameters are kept under intelligent control, to ensure the distribution network is stable, safe, reliable, and economical. Technologically, it belongs to the

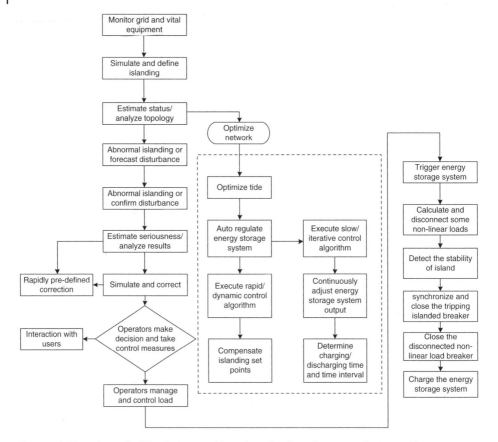

Figure 7.4 Flow chart of self-healing control based on distributed power and micro-grid.

cross-application field of power system theory, control theory, and artificial intelligence. Its control architecture is shown in Figure 7.6.

This method divides the states of the distribution network into normal operation and abnormal operation. Normal operation is further divided into hidden safety condition, dominant safety condition, economic operation state and strong operation state. It is an effective method to restore the distribution network to dominant safe operation by taking preventive actions, as shown in Figure 7.7.

Taking optimal measures, the system is able to turn from a hidden dominant safe condition to an economic operation state, as shown in Figure 7.8(a); taking strong control, the system is able to return from a dominant safe condition back to a strong operation state, as shown in Figure 7.8(b).

Correspondingly, the normal operation state is further divided into emergency, recovery, and abnormal states. The system under abnormal operating condition is able to return to normal as long as it is corrected, as shown in Figure 7.9.

Taking emergency control measures on the distribution network under emergency condition and recovery control measures on the distribution network under recovery condition, the distribution network is able to return to abnormal or normal, as shown in Figure 7.10.

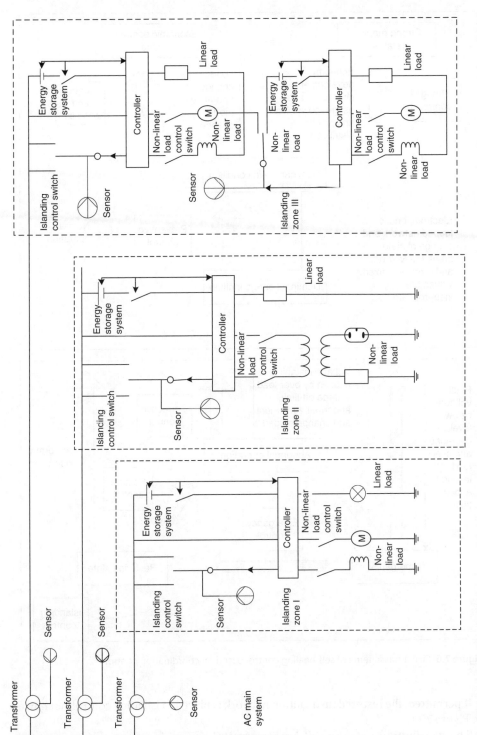

Figure 7.5 Islanding control diagram for micro-grid power supply system.

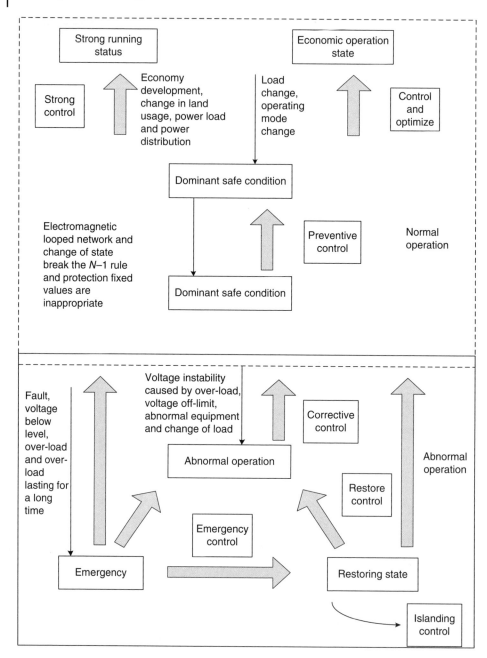

Figure 7.6 Structural scheme of self-healing control based on coordination control.

If permitted, the restored distribution network is placed in isolating control, as shown in Figure 7.11.

The distribution network self-healing control system based on the coordinated control model is able to automatically determine the current state according to the collected data and then make control decisions in a smart way according to

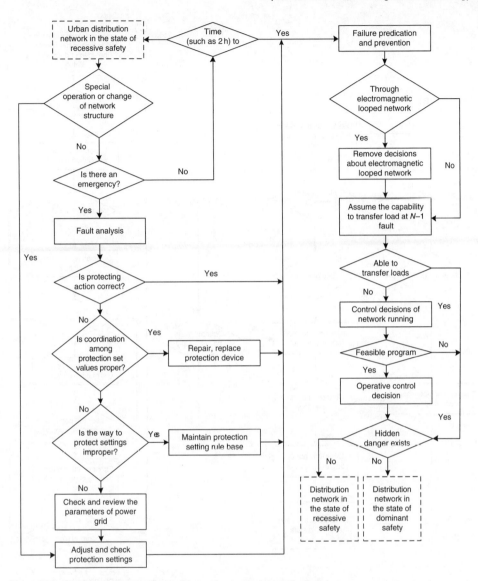

Figure 7.7 Flow chart for distribution network prevention and control.

conditions. It automatically controls relay, switchgear, automatic safety device, and automatic regulating device control, and coordinates emergency and non-emergency, normal and abnormal conditions to form a model where both dispersed and concentrated control, local and integrated control exist. This is done in order to make the distribution network survive the emergency within an expected period and restore power. At the same time, it should meet the safety restraints and be immune to the disturbance that may come with load changes. Besides, it has its own advantages, such as highly integrated and smart decisions, autonomy and coordination, redundancy and reliability, economic efficiency and adaptation.

A layered hierarchical framework for distribution network self-healing control is put forward, as shown in Figure 7.12.

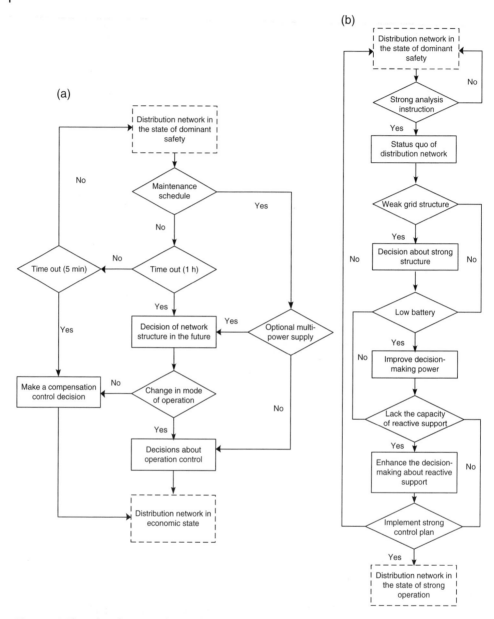

Figure 7.8 Flow chart for optimal control and strong control: (a) optimal control, (b) strong control.

This framework divides the self-healing smart distribution network system into a base layer, system layer, and advanced application layer.

1) *Base layer.* This consists of smart devices. As the physical layer of the distribution network, the more it becomes smart, the more likely it will be to implement self-healing of the distribution network.

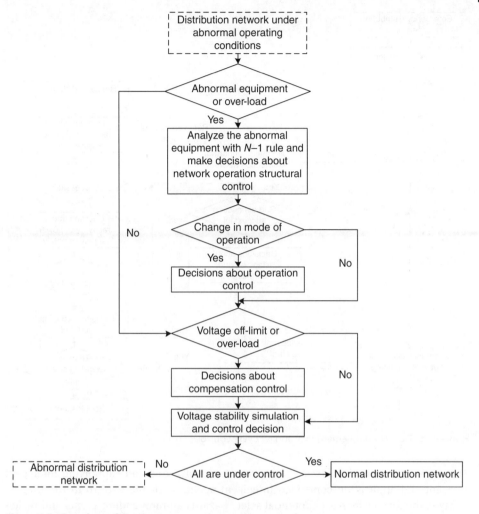

Figure 7.9 Flow chart of distribution network corrective control.

2) *System layer.* As the intermediate layer, this is composed of local smart systems, including bidirectional communication, predefined control, local supervision, data concentration, and condition preferential control. Among these, predefined control is quick to execute, since it is able to be triggered by events, including local protection, micro-grid automatic formation, and control and input of DFACTS device. Condition preferential control, set by users, has top priority and is not under the control of the advanced application layer. With the given conditions satisfied, it is quick to execute and ensures safety of the distribution network and operators, as well as a reliable power supply to important users. Local supervision is designed to monitor the regional distribution network or operation states of important devices. The system layer of data concentrated recovery should pre-process all ranges of digital quantities and analog quantities which are shared by the process layer and connect the advanced application layer with the system layer via bidirectional communication.

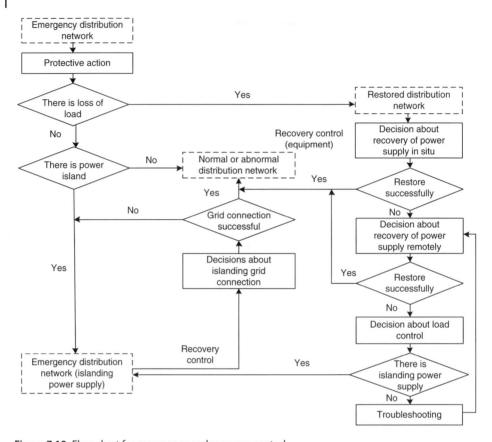

Figure 7.10 Flow chart for emergency and recovery control.

3) *Advanced application layer.* This is composed of a decision-supporting intelligent agent and application-controlled intelligent agent. With the data from the process layer, the former forecasts potential safety hazards and impending events and makes an evaluation of working conditions. Rapid simulation is an advanced software platform based on data application, where decision support for distribution networks is provided and visual results are displayed. The control intelligent agent includes formulating a control scheme, determining the final control scheme, coordinating between decentralized and integral control, improving distribution network and security control. Generally, we adopt the following control methods or combinations of them: tide control or optimization, tap adjustment, load demand management, energy storage input, distributed power generating units or renewable energy input, protecting actions, topology reconstruction, energy storage device charge and discharge control, electric vehicle charge and discharge, and offering advice on improving energy usage.

The layered frame system is able to swiftly coordinate between decentralized and integral control, optimization and security control (preventive correction, emergency recovery, repair, maintenance, and control). It can achieve both centralized control via an advanced application layer and decentralized control via a process layer. So, it acts as the fundamental organizational structure for distributed governance, wide-area coordination, condition adaptation, and attaches much importance to prevention.

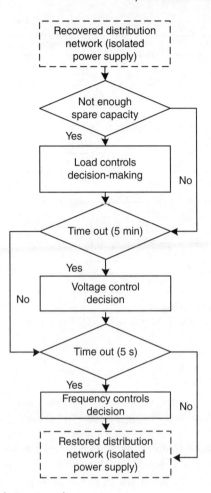

Figure 7.11 Flow chart of isolation control.

The distribution self-healing control relies on advanced information technology and strong physical grid architecture, and operates in a proper way. The self-healing control is measurement based and protection oriented, in the principle of adaption and coordination. The self-healing control stresses real-time prediction, operating evaluation, rapid simulation, and emphasizes network improvement and control security, and a combination of predefined and overall control scheme. In order to strengthen the defense capability and ensure continuous power supply, its underlying goal is to achieve uninterrupted operation and reduce losses to a minimum. The distribution network becomes a safer, stronger, and more reliable integrated facilities system.

7.3 Implementation of Distribution Network Self-healing

As mentioned previously, smart grid and self-healing control technologies are not completely new concepts, since relay protection, automatic device, and dispatching systems are part of smart grid and self-healing control technologies. The implementation of

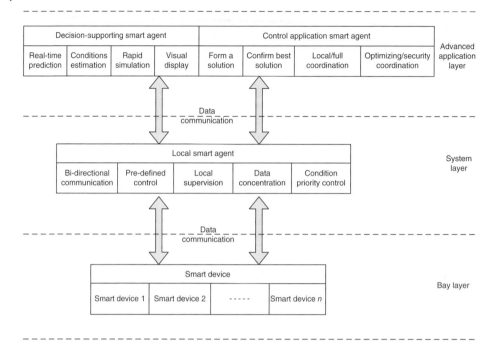

Figure 7.12 Layered frame system.

self-healing cannot be developed without the development of the above technologies. In turn, the development of self-healing control technology promotes the extension of the above technologies.

Recently, the electric power communication network has been greatly improved, and dispersed data acquisition and centralized data processing are made possible remotely in traditional relay protection, automatic device, and dispatching systems. The applied multi-point data compose the network protection and wide-area relay, security, and stability system for the large-scale grid, playing a vital role in the security defense system.

7.3.1 Self-adaptive Relay Protection Units

A self-adaptive relay device aims to allow protection to be adapted to various changes in electric power and improve the performance of the relay. An electric power system is composed of a large number of electric power units, power transmission and transformation devices, lines, and users. Its status is in constant change (including changes in user load, input/output of device, and output power of engine). Beyond that, transient or permanent failures take place occasionally. These are possibly caused by metallic short circuit or transition resistance. Hence, it is challenging for such a device to adapt to changes in electric power.

Electric power system relay protection is part of electric power automatic control in nature, and its function is to cut off the faulty device or automatically reclose the breaker to ensure normal operation. When it comes to self-adaptive protection, changes in running status and fault process should be taken into account. In the process of relay protection, information about status and fault process is locally available or obtained from a nearby substation by means of communication or by dispatching. The dispatching automation, integrated automation, and smart micro-grid together offer favorable conditions for obtaining useful information and processing it in real time. Top priority should be given to local access, since it can easily be achieved. It is comparatively complex to obtain a message from a remote terminal, with high demand for rapid transmittal of data. But if the relay functions are greatly improved, and have a suitable channel, it is also proper to realize self-adaptive protection in this way.

Broadly speaking, relay protection belongs to automatic control. The most common method of relay protection is operated by preset logic and parameters. As a result, traditional relay protection gradually becomes unable to meet the requirements for self-healing.

7.3.2 Relay Protection

7.3.2.1 Basic Requirements

Relay protection that meets the requirements of relay control should satisfy the following points:

1) support network technology and collect wide-area messages;
2) regional self-adaptive technology, automatically adapt to power grid.

Self-adaptive relay protection by nature is a control system that is able to feed back information or messages. It is compatible with self-healing technology and an actuating device for self-healing control, which is the focus of the research.

7.3.2.2 Self-adaption

Relay protection and control are adjusted and adapted as follows.

1) *System self-adaptive.* Looking back at incidents of large-scale blackouts at home and abroad, almost all are related to a malfunction in relay protection. We come to the conclusion that large-scale blackouts over the past 20 years are attributable to one common factor – inter-tripping caused by transference of tide – although the background and evolutionary processes are different. Actually, malfunctions can easily be avoided if parameters are adaptive and use wide-area information, and if a correlation factor is inserted into the relevant lines or devices.
2) *Scheme self-adaptive.* This supports protection function change, not hardware. It is very suitable for application in developing urban distribution networks. At present, some cities are faced with the problem that it is difficult for existing protection devices to meet the necessary requirements after access to new energy and bidirectional tides is introduced. If a new relay protection device is adaptive to the scheme, it will be able to cope with distributed power in the future.

Relay protection devices based on the development platform of graph-editing programming software support changes in protection function and adapt to changes in power grid, so are applicable to China's urban distribution network, as shown in Figure 7.13. An editable programming distribution network with adaptive scheme has optional configurations.

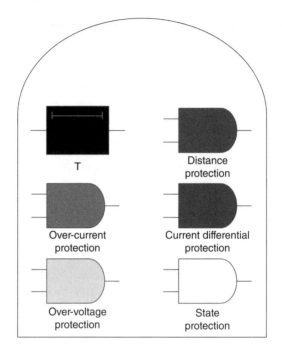

Figure 7.13 Relay protection device based on graphic logic programming software.

1) *Pilot-type protection.* This is simple and reliable and can cut off failures rapidly, so it is used as the main protection for key lines or tie lines. Such protection becomes safer and more reliable if equipped with braking. The protection is more easily realized when optical fiber is laid and carrier technology put into practice. This applies more for the HV line. As the urban economy grows, there are more and more distribution lines in urban areas, and double-circuit transmission lines on the same tower are more common, as the electricity corridors are limited in number. As a result, this type of protection is the main protection and best choice for distribution networks. A simple and self-adaptive protection designed particularly for distribution systems, in case of fault, can offer a basic guarantee of rapid self-healing control. In particular, pilot protection in the communication code of an optical fiber private network and fiber Ethernet can use a simple and effective theory of constitution.

2) *Step protection.* As backup protection, step protection is realized with protection setting values in case of communication faults. As backup protection for a line, it is generally the last line of defense. Traditionally, such protection sets, calculates, and fixes values in a fixed operating and wiring method by the step principle. When the wiring method changes, the original settings will not be applied. There will be no protection if no measure is taken. In order to isolate, cut off faults, reorganize, and self-recover, it is bound to change the original wiring method so as to maintain the best condition. The wiring and operating methods have to be reconstructed at random, depending on the location of faults. In this case, the protection needs to complete self-regulation, self-adaption, and automatic logic reorganization in the operation mode of the restructured primary system.

3) *Remote tripping protection.* To better operate the system, some lines, loads, or power sources should be cut off from a long distance. Such protection will be active in case of faults, or automatically push the system into the current state or optimum operation state when the distribution network senses something abnormal.

4) *Remote standby power sources input automatically.* Hot standby power supplies can be input to meet the requirements of the system. In the past, standby power was input in the same transformer substation or power plant, so it was a local action and need not be much considered. However, the backup power automatic input with self-adaptive scheme is designed to support the self-recovery of the global distribution network and enable the grid to enter a new optimal safe condition. So, it is operated and scheduled across the global network and is not as simple as a local action. The standby power should be started under conditions in close coordination with the status of the primary system. Automatic coordination between the primary and secondary systems is estimated by real-time information acquisition and self-healing control technology, with high demand for reliability and safety. Moreover, it is difficult to research and develop such a technique.

In conclusion, no matter what kind of protection you choose, the mode of operation is certain to change, which requires self-regulating, self-adaptive protection and a fit for the primary system in order to promote self-healing control and enter a new optimum operating state. The new relay protection makes automation truly reliable and functional. It also makes possible effective cooperation among the primary system, secondary system, network, and automation system.

7.3.3 SCADA/RTU

Supervisory control and data acquisition (SCADA) is widely used for data acquisition, supervisory control, and process control in many fields, for instance, electric power systems, feedwater systems, petroleum and chemical engineering. SCADA is also known as a telecontrol system in power systems and electrified railway. SCADA is a computer-based process control and dispatching automation system. It performs functions of data acquisition, device control, measurement, parameter adjustment, and a variety of signal alarms by monitoring and controlling field devices. The SCADA system is most widely used and developed in power systems. As the primary subsystem of the energy management system (EMS), it has the advantages of complete information, improved efficiency, current mastery of the system, acceleration of decision-making, rapid diagnosis of system faults, and is indispensable in power dispatching. It plays an important role in improving the reliability, safety, and economy of power grid operation and reducing the workload of dispatchers, realizing automation and modernization of power dispatching, and improving the efficiency and level of economic dispatching. The requirements for SCADA vary from one application area to another; consequently, the level of development for SCADA in different application fields is not completely equal.

7.3.3.1 History of SCADA
The SCADA system has always been closely related to computer technology, and has been through three generations.

The first generation of SCADA was based on specialized computer and professional operation systems, such as the SD176 system developed by Automation Research Institute for the North China power grid and the H-80 M system researched and developed by Hitachi Ltd. for China's electrified railway remote control system. The period of first generation was from the application of computers in the SCADA system until the 1970s.

The second generation of SCADA was based on a general-purpose computer in the 1980s. During the second generation, VAX and other computer and general workstations were widely used, and the operation system was the Universal Unix system. At this stage, the SCADA system built EMS in combination with economic operation analysis, automation generation control (AGC), and network analysis in the automation of power dispatching. The first and second generations of SCADA system were both based on a centralized computer system, and the system was not open, which made system maintenance, upgrading, and other interconnections difficult.

In the 1990s, with the policy of opening up, the distributed computer network and relational database enabling the interaction of EMS and SCADA was called generation 3. This generation witnessed the rapid development of SCADA and EMS, with newly developed technologies applied. During this stage, the largest investment went on electric power system automation and the power grid. China planned to channel 270 billion yuan into the renovation project of urban and rural grid in the following three years. This huge investment fund shows how much importance the Chinese government attached to electric power system automation and power grid construction.

The basic conditions for a fourth generation of SCADA/EMS system have been met. This system features Internet technology and is based on technologies such as object, neural network, and Java. It works to expand the integration of the SCADA/EMS system with other systems, and takes the requirements of both security economy and commercial operation into consideration.

7.3.3.2 Development of SCADA

The SCADA system has been improved and the technologies are in progress. As the requirements for SCADA by electric power and railway electrification systems increases, and computer technology develops, the requirements for the SCADA system are basically divided into two types.

1) *Integration of SCADA/EMS with other systems.* The SCADA system provides real-time data for electric system automation and EMS. The data are also used in simulation training and MIS. Without this, the other three systems have no data source. This can explain how connection of the SCADA system with other non-real time systems has become an important topic for research in the last decade.

2) *Remote terminal unit.* As a smart terminal, the RTU is used to measure and monitor units remotely, and mainly monitors or controls field signals and industrial equipment. Compared with the commonly seen programmable logic controller (PLC), the RTU has better communications and larger storage capacity, and is able to withstand severe climate conditions. Thanks to its perfect functions, the RTU has been widely applied in the SCADA system.

RTUs are electronics installed in the field to monitor and measure the electric equipment remotely via various sensors. An RTU will convert the status or signal into a format that is communicable in the network, and convert the data from the central computer into commands to control electrical equipment.

Supervisory control and data acquisition is a broad term for a system that supervises and controls equipment installed in the remote field. The SCADA system is widely applied in various fields, such as water, electricity, gas, alarms, communications, and safeguards, and its applications meet users' criteria and concepts of operation. The SCADA system could be as simple as connecting to one of the switches on the remote terminal through a pair of conductors, or as complex as connecting to a computer network that consists of a lot of RTUs and micro-system communications installed in a central control room. An RTU in the SCADA system can be achieved by different software and hardware, depending on the site, complexity of the system, requirements for data communication, real-time alarm and reporting, precision of measuring analog, condition monitoring, regulation of device, and switch. The RTU has the following functions.

- Collect state quantity and send remotely with priority of remote signaling.
- Send the collected data value to a remote area with optical isolation.
- Directly collect the system frequency. Measure voltage, current (active and reactive), and send the data to the remote. Calculate positive and negative power energy.
- Acquire pulse electricity and send it to the remote with optoelectronic isolation.
- Receive and execute remote controlling and back correcting.
- Automatic recovery program.
- Automatic diagnosis device (until plug-in).
- Self-governing channel.
- Supervisory channel.
- Receive and execute remote control.
- Receive and execute time correction (including GPS timing as option).
- Communicate with two or more master stations.
- Acquire sequence of events and send to the remote.
- Provide a number of digital interfaces and analog interfaces.
- Characteristics of each interface can be set from local or remote.
- Provide several communication protocols of different types; data with different protocols can be transmitted via an interface according to local and remote settings.
- Receive remote commands.
- Forward remote information from multiple substations.
- Local display function and local interface isolator.
- Support communication with spread spectrum, microwave, satellite, and carrier.
- Simultaneous operation of optional and multiple protocols, such as DL451-1991 CDT. At the same time, it supports the POLLING protocol and other international standards and codes (such as DNP3.0, SC1801, 101 and 104 protocols).
- Remote setting through telecommunication network and electric power system channel.

7.3.4 Wide-Area Measuring System and Phasor Measuring Unit

Since the existing fault recorder lacks a common time stamp at different locations, what the recorders write down is valid locally and difficult to use in the analysis of dynamic characteristics for the whole system. Therefore, it must be kept synchronous. As space expands, a wide-area electric power network is formed, but it should be operated under control. As a result, the wide-area measurement technology based on a PMU has achieved unprecedented progress. Research on PMUs and wide-area measurement systems is in full swing at home and abroad.

A power grid dynamic security detection system refers to the organic combination of synchronous PMU, high-speed digital communication device, dynamic process analysis device, and is also known as a wide-area measurement system to monitor the dynamic processes of a power grid in real time.

A GPS-based unit for measuring phase angle is designed to on-line monitor and measure the power angle of the generator and the amplitude and phase angle of bus voltage and current. Here, generators and hub substations are equipped with PMUs and connected to the control center via a communications network. The PMU corrects the time via GPS, and acquires the power angle at the same time. A time identifier is attached to the power angle, and data is sent to the control center in real time. The PMU consists of a microcomputer processor, GPS receiver, signal transformation module, and communication module.

The measurement of phasor and power angle is at the center of the PMU, in which the phasor refers to the included angle between busbar voltage and line current, relative to the reference axis, and the power angle refers to the angle between the q-axis and the reference axis. The phasor can be measured by means of zero-crossing detection and Fourier transform technique.

In the method of zero-crossing detection, the phasor synchronizes with the crystal signal of the measuring equipment by use of the pulse signal per second provided by the GPS, and generates a standard 50-Hz signal. Inside the central processing unit, a time tag is marked where there is zero crossing and the angle of a relatively standard 50-Hz signal is determined.

The Fourier transform method is based on the following principle. Assume that the frequency signal is expressed in the form of a vector:

$$x(t) = \sqrt{2}|x|\sin(\omega t + \theta)$$

$$X = \sqrt{2}|x|e^{j\theta} = X_R + X_I$$

If there are N points for sampling during a cycle, the sampling value is

$$x(t) = \sqrt{2}|x|\sin\left(\frac{2\pi}{N}k + \theta\right)$$

Through the Fourier transform algorithm, we have

$$X_R = \frac{\sqrt{2}}{N}\sum_{k=1}^{N}x(k)\cos\left(\frac{2\pi}{N}k\right) \tag{7.1}$$

$$X_I = \frac{\sqrt{2}}{N}\sum_{k=1}^{N}x(k)\sin\left(\frac{2\pi}{N}k\right) \tag{7.2}$$

If the sampling value (assume $k = 1$) corresponds to the rising edge of the GPS pulse per second, then the phasor can be calculated at the time of the second pulse at 50 Hz, according to equations (7.1) and (7.2). To calculate the phasor angle of each sample in real time, it is recommended to use a recursive Fourier algorithm.

Zero-crossing detection actually takes a sample every cycle, so the sampling period is 0.02 s and less immune to interference. The Fourier transform method that sends down the command cycle is more accurate than zero-crossing detection in the algorithm, but needs more sampling points and many calculations before the results emerge. The zero-crossing detection method is preferable if more real-time results are required.

The power angle is generally measured indirectly or directly. It can be calculated by applying a simultaneous equation and electrical quantities at the outlet of the generator. Directly, the power angle can be obtained by measuring the engine speed or location of the shaft. The latter method is more accurate, but more difficult to achieve.

7.3.4.1 WAMS System

WAMS is an integrated application system that monitors and analyzes the state of a power grid by applying synchronous phasor measurement and modern communication technologies, so that the power grid is under control and operates in real time. Figure 7.14 shows a simplified structure diagram of WAMS, including PMU-centered subsystem, wide-area communication network, master data platform system composed of data concentrators, and analysis central station system made up of a series of on-line and off-line data analysis software packages. Among them, the subsystem software includes phasor processing algorithms, integrated error estimation, compensation algorithm, and EMF electrical calculating algorithm.

7.3.4.2 PMU/WAMS and SCADA/EMS

An electric power system is generally configured with SCADA and EMS. However, what the RTU of the SCADA system can collect is limited to stable state quantities. Instead, recorders are generally responsible for collecting transient information. The PMU achieves synchronous sampling by taking advantage of GPS signals. Besides, it can directly measure the phase angle of the voltage and the voltage on buses located in a geographically vast electric power system, instead of using state estimation. The traditional method of state evaluation finds it difficult to measure the phase angle of the voltage directly, so a nonlinear least-squares method is widely used. However, this may

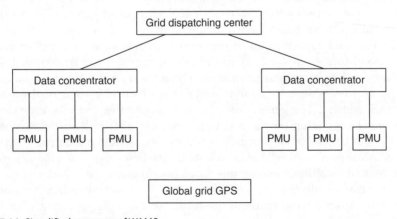

Figure 7.14 Simplified structure of WAMS.

fail to make estimations when the system scale is too large. In contrast, PMU-based WAMS is better at monitoring and controlling the dynamic processes of an electric power system. The PMU is able to measure the amplitude and phase angle of the voltage vector directly at nodes, and transmit data to the dispatching center via a high-speed communication network. If the bus in the global network is installed with a PMU, the voltage vectors at all nodes are obtained by visual observation, instead of iterative computations. Owing to the high cost of PMUs, system investment is clearly increased if all elements have a PMU installed.

The final goal of WAMS is to establish a new generation of EMS. A PMU is likely to integrate with or take the place of an RTU. Also, WAMS is likely to integrate with or take the place of SCADA. In respect of function, its goal is to improve the stationary monitoring level to a dynamic monitoring level.

7.3.4.3 Application of PMU or WAMS

The wide-area measurement technology based on PMUs has been widely applied in many areas for more than a decade.

1) *On-line low-frequency vibration analysis.* As a power grid expands and the transmission power increases, low-frequency oscillation between areas has become a serious concern for many interconnected power systems. To make WAMS discriminate it in real time, it is necessary to have a strong weak-damping oscillation component at frequency between 0.2 and 2.5 Hz, otherwise, a warning cannot be sent out. At the same time, abnormal areas should be marked on the regional graph of the power grid and platform to trigger the recording of current data. For the sake of on-line analysis, WAMS needs to provide consistent data as a real-time data platform and discriminate in which unit the vibration is taking place in the data center. In the end, conclusions should be drawn.

2) *Real value measurement without modification.* When the system is not too large, it can be used to measure real values without the need to modify the original program. Otherwise, the WAMS data increment is used to measure the redundancy so as to improve the precision. Therefore, how we evaluate the state is critical. This function is able to combine a synchronous phasor measured by WAMS with the traditional phasors of SCADA and RTU, to improve the speed and precision of state evaluation.

3) *Non-continuous and continuous control of electric power system.* The continuous control of an electric power system is similar to the protective switch for a switching reactor and capacitor. Since the output content of the controller does not correspond to the input, continuity of the input is not strictly required, so it is relatively easy to acquire measurement information via WAMS. For example, the American BPA wide-area stability and voltage control system realized WAMS feedback control. The generation tripping and reactive compensation device is performed depending on the real-time response of the electric power system. Continuous control of electric power is a similar method of control to generator excitation or governance. Thanks to WAMS, the input signal of wide-area control goes beyond local limits, so an inter-area low-frequency oscillation mode LF is the objective. In addition, the wide-area signal in some way reflects the variation in system structure and mode of operation, so the wide-area controller is self-adaptive in itself.

4) *Measurement of distance between double ends in relay protection.* The WAMS collects information with OMU as base lay, and upgrades to the dispatching center via a communication system. The PMU takes advantage of a synchronous GPS clock to measure all nodes and state quantities on the line, and convert all state quantities into a time coordinate by GPS timing. Since the GPS makes possible an accurate synchronous clock, a double-ended synchronous sampling precision distance measurement method is eventually realized. This is able to calculate the fault distance according to the voltage and current from the PMU at the ends of the line, and locate the fault accurately. This system will cover all substations when the WAMS are built all around, so there is no equipment failure in practical operations. Since the PMU is relatively high in price, how to combine the fault information system with WAMS must be taken into consideration.

5) *Wide-area protection.* The concept of wide area was first introduced and defined at the CIGRE International Conference on Large High Voltage Electric System. Depending on multiple electric power information, with rapid, reliable, and precise removal of faults, the influence of removing faults on system security is analyzed, along with the taking of control measures to improve the available capacity and/or system reliability of the transmission line.

The reason why a wide-area protection system is better than a traditional one in respect of flexibility lies in the fact that the system is able to provide multiple protections when subject to disturbance by the collection and analysis of wide-area information about the electric power system and evaluation of the states targeted by system disturbance. Wide-area protection is aimed at determining whether the system is stable after the occurrence and removal of faults. If it is prone to be unstable, it is essential to take action to secure the operation of the system.

The WAMS system has developed rapidly in recent years, in addition to advanced applications and other functions of WAMS (such as line parameter estimation, rotor angle stability predication, damp control, voltage stability control, electric power system transient analysis and control, and frequency control), and is widely used. At present, the smart grid is of great concern and there is a large space and direction of application for PMU/WAMS.

7.3.5 Smart Grid and WAMS

The smart grid is a widely recognized solution to future challenges in electric power utility, and attracts attention worldwide. The SGCC has made a commitment to innovative the world's leading strong smart grid. This also offers a great opportunity for WAMS.

Intelligent dispatching is the very center of the smart grid. According to the theory, wide-area phasor measurement is an important means to keep a large grid safe, as well as the foundation for a smart transmission network. For the time being, we embark on a smart dispatching system with aspects of WAMS and an advanced control center. So, smart control of a large-scale power grid is one of the most important links in developing the smart grid. PMU-based WAMS makes this possible. In whatever respect, this requires the promotion of WAMS and widespread use of primary substations and power plants to implement real-time observation. Efforts are being made to further study smart operation and control technology, optimize economic operation under normal power grid operation, and improve transmission capacity – all to reduce costs and achieve energy saving and efficiency improvement.

8

Pilot Project

8.1 Simulation Analysis

8.1.1 Components

The simulated analysis system consists of:

1) one set of self-sealing-based distribution automation system;
2) relay protector, ANGLI 6108G;
3) digital simulation system;
4) profile data collected by application systems.

8.1.2 Test Items

1) State conversion: safe, fault, power recovery, exception, correction, and another safe state.
2) Fault generation: during the period of maintenance or normal operation, partial faults occurring to the system can be removed by transferring loads and protecting switching settings in order to ensure continuous power supply.
3) Abnormal conditions, prevention and control (protection setting switch), correction and elimination of anomalies.
4) In case of major accidents (e.g., full power outage in Liuhe transformer substation), the power dispatching center should ensure continuous power supply to prevent the power grid from a black start (emergent control).

8.1.3 Information Flow of Simulation System

Figure 8.1 shows the information flow of the simulation system.

8.1.4 Test Results

8.1.4.1 System States
For system state items and test results, see Table 8.1.

8.1.4.2 System Management
For system management tests and results, see Table 8.2.

8.1.4.3 Self-healing Control
For self-healing control test items and results, see Table 8.3.

Self-healing Control Technology for Distribution Networks, First Edition. Xinxin Gu and Ning Jiang.
© 2017 China Electric Power Press. Published 2017 by John Wiley & Sons Singapore Pte. Ltd.

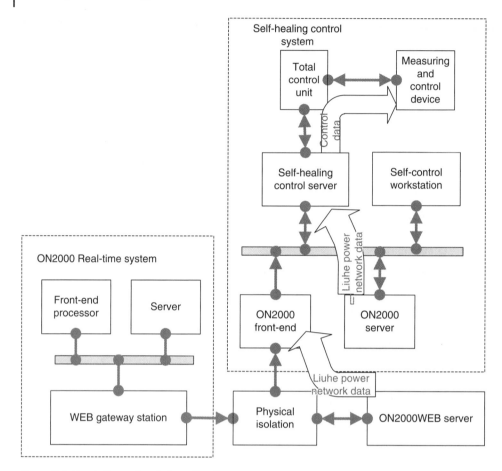

Figure 8.1 Simulation system information flow.

Table 8.1 System state test items and results

Test items	Test result	Remark	Test items	Test result	Remark
Background programs start as system starts	√		Lock	√	
System switches between real-time state and simulation state	√		Exit	√	

Note: √ indicates test is passed.

8.1.4.4   Simulation Analysis
For simulation analysis items and test results, see Table 8.4.

8.1.4.5   History Query
For test items and results of history query, see Table 8.5.

Table 8.2 System management test items and test results

Test items	Test result	Remark	Test items	Test result	Remark
Set system parameter	√		Set parameters for state estimation	√	
System establishes power grid analysis model	√		Set basic parameter of power grid	√	

Note: √ indicates test is passed.

Table 8.3 Self-healing control test items and results

Test items	Test result	Remark	Test items	Test result	Remark
Set parameters for self-healing control	√		System is able to analyze control target	√	
System can make security evaluation of power grid	√		System is able to make control strategy decision	√	

Note: √ indicates test is passed.

Table 8.4 Simulation analysis test items and test results

Test items	Test result	Remark	Test items	Test result	Remark
Save the current data into simulation database	√		System is able to simulate self-healing control	√	
Obtain data from simulation database	√		System is able to simulate prevention and control	√	
Obtain data from real-time database	√		System is able to simulate correction	√	
Adjust simulation parameters	√		System is able to restore correction	√	
Set simulation parameters	√				

Note: √ indicates test is passed.

Table 8.5 Test items and results of history query

Test items	Test result	Remark	Test items	Test result	Remark
Inquire history state of power grid	√		Inquire history alarm	√	

Note: √ indicates test is passed.

8.1.5 Simulation Cases

8.1.5.1 Simulation Case 1

When Hongma line 1 (714) is under maintenance (Hongjiangtai 714 circuit breaker and Maji 714 circuit breaker are open), failures occur to line Pingshan 311. At the time, relay protection begins to work. Circuit breaker 311 in Hongjiangtai transformer substation is disconnected, and Pingshan transformer substation falls into blackout. Largely, emergencies have been limited. The protection and relay information is uploaded to the dispatching center and the self-healing system makes the decision to restore power supply by breaking circuit breaker 304 in Pingshan substation, cutting off line Pingshan and closing the breaker 303. Besides, preventive and control measures are taken, and the test result proves that it is not necessary to switch sections of set values for the moment. At this time, a 110 kV bus is at low voltage level and the power factor of the potential transformer is too low, which are all corrected. The capacitors switch in. Figure 8.2 shows a wiring diagram for simulation case 1.

8.1.5.2 Simulation Case 2

Increase the load of Masi 306 and suppose that 714 Hongma line 1 is faulty (i.e., breaker 714 of Hongjiangtai transformer substation and breaker 714 of Maji are open) and

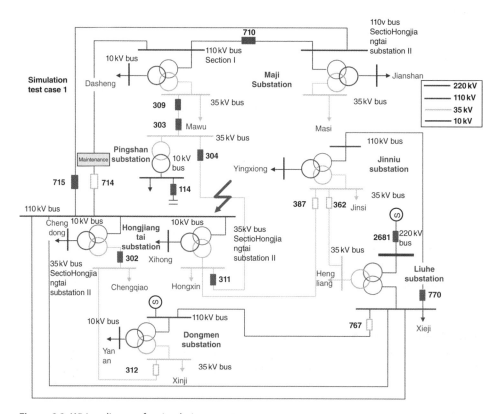

Figure 8.2 Wiring diagram for simulation case 1.

crosses 715 Hongma line 2. In this case, to solve the problem, we can close the 303 breaker in Pingshan transformer substation, open the 710 breaker in Maji, and transfer the load of 306 Mawu to Pingshan transformer substation. At the same time, it is necessary to take preventive and control measures by switching a section of fixed values in Hongjiangtai 311. The low voltage of the 10 kV bus in Pingshan transformer substation is corrected by means of adjusting the gear on the main transformer. After the recovery of Masi's load, open the breaker 303 in Pingshan transformer substation and close the breaker 710 in Maji transformer substation. Then, the voltage of the 10 kV bus in Pingshan transformer substation must be overly high. This can be corrected by adjusting the gear on the potential transformer, and preventive and control measures must be taken to switch back to a section of fixed value 311. Figure 8.3 shows the wiring diagram of simulation case 2.

8.1.5.3 Simulation Case 3

The background for this case is that cut-off of supply occurs to the global Liuhe transformer substation (power circuit breaker 2681 breaking). At this time, it is necessary to launch emergency control over the urban network. By means of load shedding, voltage regulation, and other measures, the 110 kV bus that is supposed to be connected to the

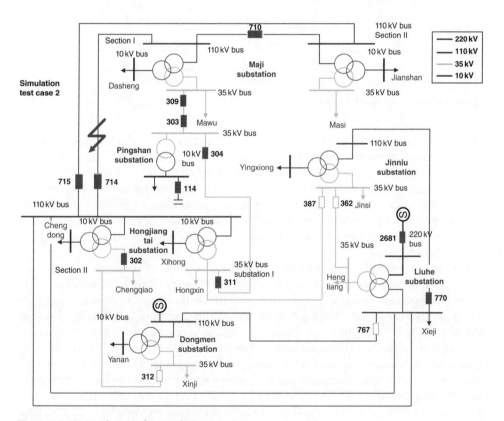

Figure 8.3 Wiring diagram for simulation case 2.

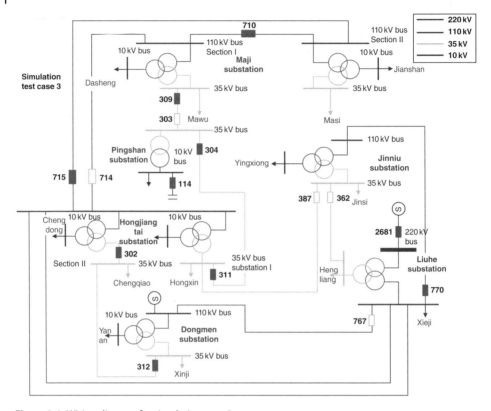

Figure 8.4 Wiring diagram for simulation case 3.

Dongmen transformer substation is made to supply part of the power to the Hongjiangtai transformer substation, ensuring power supply to Chengqiao, the original power supply to Liuhe dispatching and directing center. This can be done by opening the circuit breaker 302 at the LV side of Hongjiangtai transformer substation and making the breaker 312 of Dongmen transformer substation. In order to ensure continuous power supply to Chengqiao, cyclical load shedding of the remaining loads of Hongjiangtai transformer substation is necessary. Figure 8.4 shows the wiring diagram for simulation case 3.

To carry out a simulation test on the above cases, the states of the breakers and measured data have to be sent to a monitoring system. These breakers include: breaker 714 in Maji substation (Hongma line 1), breaker 710 (section breaker), breaker 309 (line Ma-ping 311), breaker 303 in Pingshan substation (line Ma-ping 311), breaker 304 (line Pingshan 311), breaker 106 (load), breaker 114 (capacitor); breaker 714 of Hongjiangtai transformer substation (Hongma line 1), breaker 715 (Hongma line 2), breaker 302 (potential transformer), breaker 308 (Chengqiao load), breaker 311 (line Pingshan 311), breaker 312 in Dongmen substation (line Donghong); breaker 387 in Jinniu substation (line Jinhong), breaker 362 (line Babai), breaker 767 in Liuhe substation (line Liudong), breaker 2681 (power source). Breakers at the above points are connected to DSA1000 programmable logic protection and measuring/control devices.

8.2 Pilot Application

8.2.1 Requirements for Pilot Power Grid

In the demonstration area, a 220 kV transformer substation has been at the center of the city and supplies power to users via a 110 kV, 35 kV power grid, distributed via a 10 kV power grid, respectively. In the demonstration area, an urban HV distribution network system is made up of 110 kV substations and 10 kV buses, including power and distributed power direct access to the voltage level.

The distribution network for the demonstration project has the following characteristics. The main network, in contrast to the distribution network, amounts to an infinite bus system. Owing to the short distance between the distributed power supplies within the distribution network, the power angle always remains stable. Voltage stability becomes a concern, since the loads for domestic appliances increase year by year; when the distribution network is split from a large-scale system, the huge disturbance that comes with it may lead to an imbalance between power supply and loading power, and problems of frequency stability. For the distribution network, it is best to perform a partial closing loop operation, which is relevant to the structure of the power grid and load level. For this reason, it should be taken into account in the analysis of safe, reliable, and economic operation.

In the theory of self-healing control, the distribution network in a certain state should be put under control until it achieves a better state in order to ensure stability, security, reliability, and economic operation of the network. In the system, such method of control is collectively referred to as self-healing control. It is defined as a means of control by running status that enables the system to transfer from the current state to a better state, survive emergencies, and quickly restore power. At runtime, the system in the state should meet security constraints as well as high economic efficiency. It should also be immune to interference caused by load changes.

To improve the security and reliability of the demonstration network, emphasis should be put on the load transferring capacity, particularly MV and LV of the distribution network, and reducing the losses caused by power failure. It is required that the automation system in the distribution network should be adapted to accommodate swift changes of operating mode. Beyond that, a secondary system provides backup for the primary system, instead of retarding it. Therefore, the relay protection and control system is expected to adapt to change. At the same time, the following considerations should be taken into account.

1) Since distributed power features small capacity, low system voltage, and frequent input/output, these characteristics have a great impact on the relay protection and control system at access points. Efforts should be made to work out a secondary protection and control strategy accordingly.
2) Considering both economics and practicality, the idea of combining real-time monitoring and mathematical statistics to optimize the location of remote control and telemetering brings about benefits.
3) In demonstration mode, use the integrated on-line monitoring terminal and system of the primary device to explore the method of predicting faults that may occur in the distribution network. In the demonstration area, substations and power sources

that may transfer loads to each other are able to achieve internal load balance, coordination, and transference. Nanjing urban network system has been chosen for the demonstration engineering.

8.2.2 Contents of Demonstration Project

1) Research theories about self-healing control technology integrated with practice:
   - data extraction, flow chart building, and mathematical model;
   - simulation method of demonstration project;
   - test method applied to simulation system;
   - data access via scheduling system, field test, and scheme validation.
2) Check relay protection and control strategy as well as the consistency and validity of advanced applications in self-healing strategy.
3) Install protection and measuring/control devices at the site and use them to collect data and conduct tests.
4) Equip primary device on-line integrated monitoring terminal and system to explore ways to predict the potential faults in an urban power grid.
5) Build a communication system according to the existing resources and substation to ensure the dispatching center is equipped with communication channels.

8.2.3 Distribution Network of Pilot Project

1) Transformer substations and potential transformers
   11 sets of 110 kV substations, 18 sets of main transformers, with capacity of 522.5 MV·A.
   15 sets of 35 kV transformer substations, 30 sets of main transformers, with capacity of 202.25 MV·A.
   3153 sets of 10 kV public distribution transformers, with capacity of 475.352 MV·A.
   Also governs No. 4 and 5 main transformers of Substation A, with capacity of 31,500 kV·A, respectively.
   No. 1 and 2 main transformers of Substation B, with capacities of 120,000 kV·A and 180,000 kV·A.
   No. 1 and 2 main transformers of Substation C, with capacities of 120,000 kV·A and 180,000 kV·A.
2) Overhead lines and cables
   One 220 kV overhead line, 70.433 km in length.
   23 110 kV overhead lines, 209.058 km in length; 21 110 kV cables, 19.623 km in length.
   27 35 kV overhead lines, 343.773 km in length; 482 35 kV cables, 103 km in length.
   162 10 kV overhead lines, 2439 km in length; 482 10 kV cables, 103 km in length.
   There are two other 110 kV Internet lines in Liuhe region, with a length of 5.65 km, and Liuhe Power Supply Company will be responsible for operations and maintenance.
3) User power plant and installed capacity
   Power plant A: six 60,000 kW substations.
   Power plant A1: three 41,000 kW substations and one 6000 kW substation.
   Company B: two 25,000 kW substations and one 6000 kW transformer substation.
   Company C: two 25,000 kW substations, one 7000 kW, 9000 kW, 10,000 kW, and 12,000 kW substation, respectively.

Company D: two 50,000 kW substations.

Company E: two 6000 kW substations.

The transformer substations total 30, with total installed capacity of 850,000 kW.

4) Research and application results of pilot engineering project

Interim deliverables can be put into trial operation, combined with engineering, according to the project schedule and progress. The research and application results of trial projects include:

- On-line logic programmable relay and measuring/control devices in accordance with IEC 61131-1.
- System for collecting dynamic information of electrical equipment.
- Advanced application software for self-healing control technology set up in dispatching system of Liuhe district.

Sensors and measuring terminals installed on the extended switch cabinet in Jinniushan, Liuhe district make on-line monitoring possible. The Hongjiangtai substation has been equipped with protective cabinets on both sides, installed with programmable logic protective device and intended for collecting parameters related to the 35 kV tie line.

The dispatching system for Liuhe district uses an ON2000 dispatching automation system, which ensures successful implementation of the project. The mature SCADA system carries on its role in the project.

Figure 8.5 shows the installation and test site for the on-line supervision system of the breaker.

5) Demonstration of distribution network wiring

The urban area of Liuhe and urban/rural conjunction are given as a demonstration plot for the distribution network. The indefinite system power is by 220 kV bus in Liuhe transformer substation. Hongjiangtai transformer substation will serve as a

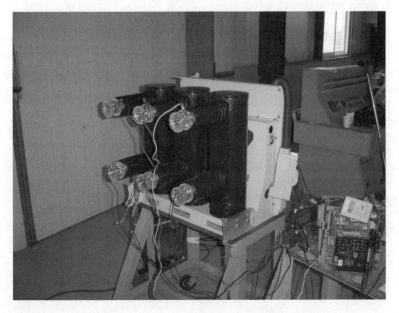

Figure 8.5 Installation and test site for on-line supervision system of breaker.

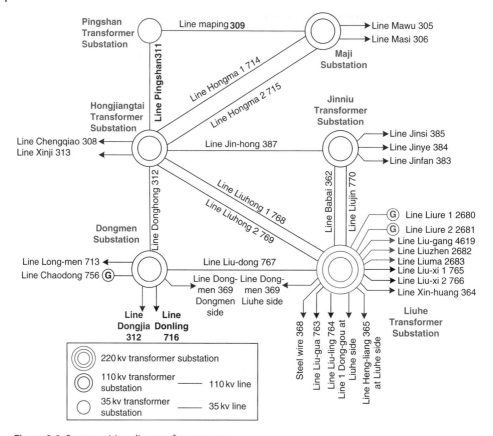

Figure 8.6 System wiring diagram for test area.

key place to ensure power supply, and line Cheng-qiao will supply power to the dispatching and directing system of Liuhe district. In addition, financial, hospital, and emergency centers in urban areas are also key units for ensuring power supply. See Figure 8.6 for the system wiring diagram of the test area.

6) Parameters and system profile information
   - Line data: see Table 8.6.
   - Reactive compensation capacity: see Table 8.7.
   - Typical profile data: see Table 8.8.
   - Load data of Pingshan substation: see Table 8.9.
   - Load data of Hongjiangtai substation: see Table 8.10.
   - Loads of Dongmen transformer substation: see Table 8.11.
   - Data of loads in Jinniu substation: see Table 8.12.
   - Load data in Maji substation: see Table 8.13.
   - Load data of Liuhe substation: see Table 8.14.

Table 8.6 Line data

Serial number	Head-end plant/station	Terminal station	Line name	Length (km)
1	Maji transformer substation	Pingshan transformer station	Line Ma-ping 309 (303)	11.528
2	Hongjiangtai transformer substation	Maji transformer substation	Line Hongma 1 714	19.989
3	Hongjiangtai transformer substation	Maji transformer substation	Line Hongma 2 715	19.807
4	Hongjiangtai transformer substation	Pingshan transformer substation	Line Pingshan 311 (304)	9.399
5	Dongmen transformer substation	Hongjiangtai transformer substation	Line Donghong 312	2.669
6	Jinniu transformer substation	Hongjiangtai transformer substation	Line Jinhong 387	9.377
7	Liuhe transformer substation	Dongmen transformer substation	Line Liudong 767	10.871
8	Liuhe transformer substation	Hongjiangtai transformer substation	Line Liuhong 1 768	7.245
9	Liuhe transformer substation	Hongjiangtai transformer substation	Line Liuhong 2 769	7.245
10	Liuhe transformer substation	Jinniu transformer substation	Line Liujin 770	7.078
11	Liuhe transformer substation	Jinniu transformer substation	Line Babai 362	5.132

Serial number	Resistance (p.u.)	Reactance (p.u.)	Conductivity (p.u.)	Susceptance (p.u.)	Charging power (Mvar)
1	0.2336	0.3279	1.441	−2.0227	0.0424
2	0.0408	0.0617	7.4629	−11.2773	0.7256
3	0.0404	0.0611	7.5315	−11.3809	0.719
4	0.3175	0.2794	1.7752	−1.56215	0.0345
5	0.0541	0.0759	6.2239	−8.73647	0.0098
6	0.19	0.2668	1.7715	−2.48669	0.0345
7	0.014	0.0324	11.2315	−26.0306	0.3946
8	0.0093	0.0216	16.8526	−39.0584	0.263
9	0.0093	0.0216	16.8526	−39.0584	0.263
10	0.0091	0.0211	17.2502	−39.98	0.2569
11	0.104	0.146	3.2369	−4.54358	0.0189

Note: Reference value is 100 MVA; rated value = reference value.

Table 8.7 Capacity of reactive compensation

Name of power plant	Capacity of capacitor (Mvar)	Circuit breaker	Voltage class (kV)	Remark
Liuhe substation	7.2	305	35	
	7.2	306		
Dongmen substation	3.6	105	10	Section I
	3.6	205		Section II
Jinniu substation	3.6	103 A	10	Section I
	3.6	104 A		Section II
	1.8	103 B		Section I
	1.8	104 B		Section II
Hongjiangtai substation	2.4	115	10	Section I
	0.9	118		Section II
Maji substation	2.4	105	10	Section I
Pingshan substation	2.4	114	10	

Table 8.8 Typical profile data collected by system

Category	Name of line	Active tide (MW)	Reactive tide (Mvar)	Voltage (kV)	Current (A)	Measuring point
Line	Line Maping 309 (303)	0	0	35.54	0.59	Maji substation
	714 Hongma Line 1	7.93	3.17	111.6	43.51	Maji substation
	714 Hongma Line 1	7.79	2.21	112.6	41.14	Hongjiangtai substation
	715 Hongma Line 2	2.64	1.36	112	15.29	Maji substation
	715 Hongma Line 2	2.62	0.59	112.57	13.89	Hongjiangtai substation
	311 Line Pingshan (304)	1.86	1.022	35.07	34.94	Hongjiangtai substation
	Line Pingshan 311 (304)	1.83	0.87	34.83	33.6	Pingshan substation
	Ling Donghong 312	0	0.057	35.58	0.94	Hongjiangtai substation
	Line Donghong 312	0	0	35.97	0	Dongmen substation
	Line Jinhong 387	0.01	0.056	35.07	0.94	Hongjiangtai substation
	Line Jinhong 387	0	0	36.82	0	Jinniu substation
	Line Liudong 767	5	−0.1	112.7	25.62	Liuhe substation

Table 8.8 (Continued)

Category	Name of line	Active tide (MW)	Reactive tide (Mvar)	Voltage (kV)	Current (A)	Measuring point
	Line Liudong 767	−4.69	0	112.65	24.62	Dongmen substation
	768 Liuhong Line 1	−16.47	−4.75	112.57	87.72	Hongjiangtai substation
	769 Liuhong Line 2	−8.82	−0.34	112.57	46.24	Hongjiangtai substation
	768 Liuhong Line 1	16.9	4.7	112.7	89.86	Liuhe substation
	769 Liuhong Line 2	8.8	0.2	112.7	45.09	Liuhe substation
	702 Line Liujin	11.88	3.3	112.42	66	Jinniu substation
	770 Line Liujin	13.1	2.5	112.7	68.32	Liuhe substation
	362 Line Babai	0	0	36.82	0	Jinniu substation
	362 Line Babai	0	0	36.4	0.6	Liuhe substation
Power supply	2Y23 Liushan Line 1	−9.2	−6.0	228.7	27.73	Liuhe substation
	2Y24 Liushan Line 2	−9.6	−5.8	229.2	28.25	Liuhe substation
	756 Line Chaodong	−12.12	−0.89	119.93	58.21	Dongmen substation
Equivalent load	368 Line Gangchang	0	0.3	36.4	17.9	Liuhe substation
	369 Line Dongmen	0	0	36.4	0	Liuhe substation
	763 Line Liugua	17.4	3.2	112.7	90.63	Liuhe substation
	764 Line Liuling	6.7	−0.3	112.7	34.36	Liuhe substation
	366 Donggou Line 1	1.5	0.8	36.4	26.96	Liuhe substation
	365 Line Hengliang	0	0	36.4	82.3	Liuhe substation
	364 Line Xinhuang	1.7	1	36.4	31.28	Liuhe substation
	765 Liuxi Line 1	0.1	0.1	112.7	0.7245	Liuhe substation
	766 Liuxi Line 2	0.1	0	112.7	0.5123	Liuhe substation
	2683 Line Liuma	−27.2	3.8	228.7	69.33	Liuhe substation
	2682 Line Liuzhen	−45.1	2	228.7	113.97	Liuhe substation

(*Continued*)

Table 8.8 (Continued)

Category	Name of line	Active tide (MW)	Reactive tide (Mvar)	Voltage (kV)	Current (A)	Measuring point
	619 Line Liugang	0	0	0	0	Liuhe substation
	369 Line Dongmen	0	0	36.45	0	Dongmen substation
	716 Line Dongling	2.9	0.03	112.65	13.89	Dongmen substation
	304 Line Dongjia	0.15	0.104	35.97	2.93	Dongmen substation
	713 Line Dongmen	0	0	119.93	0.35	Dongmen substation
	385 Line Jinsi	1.25	1.81	36.82	34.46	Jinniu substation
	384 Line Jinye	3.94	1.95	36.82	68.91	Jinniu substation
	383 Line Jinfan	0.61	0.28	36.82	10.55	Jinniu substation
	308 Line Chengqiao	3.85	1.22	35.58	63.31	Hongjiangtai substation
	313 Line Hongxin	0	0	35.07	0	Hongjiangtai substation
	305 Line Mawu	4.61	1.7178	35.54	79.92	Maji substation
	306 Line Masi	0	0	35.52	0.7	Maji substation

Table 8.9 Loads of Pingshan substation

Name	Active power (MW)	Reactive power (Mvar)	Voltage (kV)	Current (A)
Capacitance compensation	0	0	10.46	0
Ling Hongmiao	0.34	0.2199	10.46	22.35
Line Linchang	0.64	0.2198	10.46	37.35
Line Lianshan	0.25	0.1133	10.46	15.15
Line Xinhua	0.62	0.3769	10.46	40.05
Total	**1.85**	**0.9299**		

Table 8.10 Loads of Hongjiangtai substation

Name	Active power (MW)	Reactive power (Mvar)	Voltage (kV)	Current (A)
Section II capacitance compensation	0	−0.81	10.47	44.39
Railway station	0	0.024	10.47	1.32
Yifeng station	0	0.0096	10.47	0.53
Chengdong station	0.251	0.021	10.47	13.89
Longhai station	1.022	0.443	10.47	61.43
Line Hongdian	0.306	0.188	10.47	19.81
Line Nongke	0.732	0.37	10.47	45.25
Line Beimen	1.582	0.595	10.54	92.59
Line Xihong	0.835	0.386	10.54	50.40
Line Chengxi	1.783	0.974	10.54	111.28
Line Chengbei	1.124	0.491	10.54	67.18
Bypass	0	0.0064	10.54	0.35
Line Hongling	0.465	0.186	10.54	27.43
Line Honglong	0.962	0.564	10.54	61.08
Section I capacitance compensation	0	−2.23	10.54	120.69
Total	**9.062**	**1.218**		

Table 8.11 Loads of Dongmen substation

Name	Active power (MW)	Reactive power (Mvar)	Voltage (kV)	Current (A)
Line Xiangtang	2.434	0.3	10.50	134.84
Tangcheng Line 2	0.569	0.273	10.50	34.70
Capacitance compensation	0	−3.35	10.50	184.07
Jinning Line 2	0.614	0.239	10.50	36.23
Spare line	0	0	10.50	0
Line Yefu	0.975	0.192	10.50	54.64
Line Yanan	0.102	0.084	10.50	7.27
Line Renhe	0.531	0.801	10.33	53.70
Line Gongyuan	0.691	0.116	10.33	39.16
Capacitance compensation	0	−1.14	10.33	64.02
Line Chaotian	0	0	10.33	0
Jinning Line 1	0.411	0.245	10.33	26.73
Tangchen Line 1	0	0	10.33	0
Line Nanmen	2.363	0.88	10.33	140.93
Total	**8.69**	**−1.36**		

Table 8.12 Jinniu substation load data

Name	Active power (MW)	Reactive power (Mvar)	Voltage (kV)	Current (A)
105 A	0	−3.429	10.52	190.74
104 B	0	0	10.52	0
Line Yingxiong	0.11	0.0214	10.52	6.15
Line Jiancai	0.992	0.7053	10.52	66.8
Line Datang	1.215	0.4363	10.52	70.85
Line Baidu	0.092	0.0374	10.52	5.45
Standby	0	0	10.52	0
Standby	0	0	10.52	0
Standby	0	0	10.52	0
Line Bajie	0.721	0.4601	10.52	46.94
Spare	0	0	10.52	0
Line Shuangjing	1.193	0.7505	10.52	77.35
Line Changshan	0.82	0.5934	10.52	55.55
Line Sunlin	0.834	0.4512	10.52	52.04
103 A	0	0	10.52	0
103 B	0	0	10.52	0
Total	**5.977**	**0.0266**		

Table 8.13 Load data of Maji substation

Description	Active power (MW)	Reactive power (Mvar)	Voltage (kV)	Current (A)
Line Jianshan	0.265	0.183	10.31	18.02
Line Huanggang	0.935	1.0317	10.31	77.97
Line Yuwang	0.114	0.0563	10.31	7.12
Line Dasheng	0.511	0.364	10.32	35.10
Line Maji	0.591	0.305	10.32	37.20
Capacitance compensation	0	−1.07	10.32	60.24
Total	**2.416**	**0.87**		

Table 8.14 Load data of Liuhe substation

Description	Active power (MW)	Reactive power (Mvar)	Voltage (kV)	Current (A)
No. 1 capacitance compensation	0	0	36.40	0
No. 2 capacitance compensation	0	−6.0	36.40	95.5
Potential transformer	0	0	36.40	0
Bai-bai Line			36.40	0.6
Dong-men Line	0	0	36.40	0
Dong-gou Line	1.5	0.8	36.40	26.96
Xin-huang Line	1.7	1.0	36.40	31.28
Heng-Liang Line			36.40	82.3
Steel mill	1.088	0.3	36.40	17.9
Liu-xi Line 1	0.1	0.1	112.8	0.7238
Liu-xi Line 2	0.1	0	112.8	0.5118
Liu-dong Line	5.0	−0.1	112.8	25.60
Liu-hong Line 1	16.9	4.7	112.7	89.86
Liu-jin Line	13.1	2.5	112.7	68.32
Liu-hong Line 2	8.8	0.2	112.8	45.05
Liu-ling Line	6.7	−0.3	112.7	34.36
Liu-gua Line	17.4	3.2	112.8	90.55
Line Xie-ji	1.5	0.9	112.8	8.95
Total				

9

Development Progress of Smart Grid in the World

9.1 Introduction

Owing to the joint efforts of countries from all around the world in research, pilot application, and capital input, since its publication three years ago, smart grid has developed rapidly in various fields.

The technology methods may vary from state to state, but all of them have the same basic goal as follows: optimizing the distribution of resources, ensuring the stability, reliability, and economy of power supply, meeting environmental protection constraints, ensuring power quality and market orientation in order to provide consumers with a reliable, economic, clean, and interactive power supply and value-added service.

As power electronics, communication technology, automatic control technology, and big data technology develop, the concept of smart grid has been further developed and it has focused on the value of information flow and integration of energy flow since 2009 when the SGCC launched a program to study and build a smart grid.

The Chinese smart grid follows three steps, and has advanced the construction and implementation of a strong smart grid in a scientific and orderly way. Over the years, there have been up to 311 smart grid pilot programs of 32 types in place. The Chinese smart grid pilot projects cover the largest area, with the largest scale of construction and highest rate of speed in the world. 64 of them are station-level power grid projects, such as the National Development and Reform Commission (NDRC), Ministry of Science and Technology, and National Standard Committee. In addition, great successes have been achieved in the technologies of informatization, automation, and interaction from 32 key projects in 10 fields, such as new energy grid connection, monitoring the states of transmission and transformation equipment, smart substation, distribution network automation, electricity consumption gathering, smart power utilization service, EV charging, switching service network, and dispatching control system.

Despite the economic downturn, developed countries in the world still prioritize smart grid as an important national development strategy, and advance its construction with the guidance and support of government and through the joint efforts and collaboration of government and enterprises. Smart grid is of vital importance to reach the commanding heights of a future low-carbon economy.

Self-healing Control Technology for Distribution Networks, First Edition. Xinxin Gu and Ning Jiang.
© 2017 China Electric Power Press. Published 2017 by John Wiley & Sons Singapore Pte. Ltd.

9.2 Current Situation of Chinese Smart Grid: China's National Strategy

9.2.1 Distribution Network Automation

SGCC initiated the first urban power distribution automation pilot projects in central zones (or parks) in Peking, Hangzhou, Yinchuan, and Xiamen in August 2009. These pilot projects were designed to build a smart distribution network with features of self-healing, user interaction, efficient operation, customized power, and distributed generation flexible connection by taking a proper power distribution automatic technical configuration solution according to reliability demands. Based on the first-phase construction of the first pilot urban distribution automation program, the second-phase construction put a high priority on extending distributed power access technology supports, perfecting the construction of a supporting platform for advanced applications of power distribution network and scheduling control integrated technology, and making integrated management come true.

9.2.2 Standards Release

SGCC released and revised its *Smart Grid Technology Standard System Plan* in June 2010 and October 2011, respectively, covering eight professional branches, 26 fields of technology, and 92 standard series in total. By the end of 2012, SGCC had released 220 smart grid enterprise standards and established 75 smart grid industrial standards, 26 Chinese national standards, and seven international standards.

As the *12th Five-Year Plan Outline* released in 2011 prescribes, it is required to boost the construction of a smart power grid relying on information, control, power storage technologies, and the like.

According to the *12th Five-Year Plan of Smart Grid Major Scientific and Technological Industrialization Project* released by the Ministry of Science and Technology in May 2012, it has put priority on the large-scale intermittence new energy grid technology, all power system technologies that support the development of electric vehicles, large-scale energy storage systems, smart power utilization technology, large power grid intelligent operation and control, smart transmission and transformation technology and equipment, power grid information and communication technology, flexible transmission and transformation technology and equipment, and smart grid integrated technology.

China Electricity Council, together with SGCC, held a kick-off meeting on May 31, 2013, which marked the start of a comprehensive smart grid standardization program. The program should transfer technological achievements – especially independent innovative achievements – into standard achievements in a new energy grid, smart transformer substation, intelligent dispatching, charging and switching of electric vehicles, relying on the efforts of government, industry, and enterprise, and apply these achievements in the standard system of smart grid professional technology, consistent with national standards, industry standards, and enterprise standards.

9.2.3 Research and Demonstration

The Mondon Chenqi Herhongde solar and wind storage complementary micro-grid pilot project took place in June 2012. The Old Barag Banner immigration village is a new living place for herdsmen from Harigantuza, Huhenuer Town, which is mostly heavy sand. The village is an eco-friendly immigration village built in 2010 by the Inner Mongolian Autonomous Region. There are 100 herdsmen, one milk station, one tap water station, and one village committee office, covering an area of 120,000 square meters. The village has access to a power grid via a 35 kV light-duty wire, and the system loads are borne by new energy to ensure the power supply in case of power failure. The Chenqi micro-grid pilot project was provided with 110 kWp photovoltaic, HY-20 kW wind turbine, HY-30 kW wind turbine, 42 kWh lithium, and PCS. As distributed power sources, the 30 kW wind turbine and 80 kWp photovoltaic panel mainly supply power for grid-connected power generation and Line 1 for villagers. As part of the new energy within the micro-grid, the 20 kW wind turbine and 30 kW photovoltaic panel mainly supply power to Line 1 of the immigration village, with a lithium battery charged via PCS. The construction of these pilot projects has addressed technical problems in distributed power, energy storage, and micro-grid connection and control, and explored the access and construction mode of distributed power generation, energy storage, and micro-grids in rural power grids.

11 subprojects of comprehensive demonstration projects in Yangzhou Economic Development Zone were completed in December 2012; the third smart grid demonstration project following Shanghai World Expo, Sino-Singapore Tianjin Eco-city. The deliverables of the project include distribution network automation, user information collection, electric vehicle charging facilities, PV grid connection, micro-grid operation and control, and display visualization application.

Zhejiang built its first integrated smart grid demonstration project in the Jinghu District of Shaoxing City in May 2013. The project brought together China's latest achievements in the fields of power generation, transmission, distribution, utilization, and dispatching, and the full-view monitoring and smart marketing information platform helped to collect information from and monitor HV and LV devices.

Delta Green Tech created a distributive visual control system (DVCS) in 2013, and its appearance at the 2013 China International Smart Grid Technology and Equipment Exhibition introduced a whole new dimension to the techniques of display and control. In June 2013, the starting area of Sino-Singapore Tianjin Eco-city built 12 smart grid subprojects. Full coverage of power optical fiber will be phased in to provide technical convenience for power utilization and support the development of smart cities.

The construction scheme of key laboratory research for the Jiangsu smart grid distribution and utilization technologies was finally passed by experts, through demonstration in July 2013. The laboratory will develop a series of key technologies required in power grid distribution and utilization, including integrated intelligent equipment, efficient PV power generation, and high-speed two-way interactive communication.

The construction plan for the Lanzhou new area smart grid project of Gansu Electric Power Company passed the assessment of SGCC experts in August 2013. The project was Gansu's first large-scale integrated smart grid project, and will contribute positively to the construction of a smart city in Lanzhou.

The *863 Program Grid Key Technology Research & Development (Phase I) 19th Project* to "enhance the security and stability of power grid and flexible control of operating efficiency" was adopted in October 2013. Research results will be applied to the mid-China power grid and the Ximeng transmission demonstration project, and cross-district AC/DC coordination and control/assistant decision systems will be constructed in the central China power grid, realizing wide-ranging security, stability, control, and coordination for the first time. While the simulation test platform for the demonstration project was built in a state laboratory for grid security and energy saving, it has improved the simulation ability of the mounting controller and AC/DC power coordination and control. Relying on the central China power grid demonstration project, the Ximeng transmission project will test the roles and effects of damping control by cross-district DC coordination and control, with a large-scale thermal power fleet delivery system under laboratory environments to support the construction of demonstration projects.

The China Southern Power Grid Company and China Mobile renewed a strategic cooperation framework agreement in November 2013 to construct a smart grid by virtue of 4G. China Mobile will provide China Southern Power Grid Company with information solutions for office, scheduling, and management, as well as mobile office and other information applications and services, and cooperate on constructing and improving the communication management and quick response of remote meter-reading system terminals. The two sides will integrate their advantages in terms of network resources and develop extensive cooperation in TD-LTE communication technology, transport network, voice service, marketing automation, information technology, and so on.

The subproject of national project 863 entitled *Smart Grid Optimized Dispatching Key Devices Development and Application* passed appraisal of scientific/technical achievements by the Chinese Society of Electrical Engineering (CSEE) in November 2013. The project studied a distribution network that included distributed power or diversified loads and other new elements, and mainly explored an optimization schedule model of smart grid using breakthrough key technology, developing key equipment to underpin the optimization and scheduling of smart grid, developing an optimized scheduling system, and carrying out demonstration application in order to achieve efficient operation.

Yetang 110 kV substation, a new generation of smart substation designed by Shanghai Electric Power Design Institute Limited, was put into operation in December 2013. Yetang substation is one of six new intelligent substation demonstration projects initiated by SGCC. Compared with conventional intelligent substations, the new generation of smart Yetang substations is characterized by integrated smart devices and integrated business systems, using integrated equipment, networks, and system technology architecture, shifting from professional design to integration design, from the 'intelligentization' of primary devices to primary smart equipment.

In addition, Shanghai issued the *Shanghai Initiated Action Program for Development of Smart Grid Industry (2010–2012)* and Jiangsu issued *Jiangsu Smart Grid Industry Development Special Plan Outline (2009–2012)*, becoming pioneers in the local development of smart grid. More than 193 cities, including Peking, Shanghai, Guangzhou, Shenzhen, Hangzhou, and Nanjing, are listed as pilot smart cities, and energy-saving and new-energy vehicles have been introduced in Shanghai, Shenzhen, Guangzhou, and Hefei as part of the pilot work, driving the application of smart grid in China.

China stands at the same starting line with European and American countries in respect of studying and constructing smart grid. Many international standards have been prepared by Chinese research institutes, changing the fact that countries or enterprises outside China used to prepare international standards for electric power systems.

The *Strong Smart Grid Technology Standard System Planning* clearly specifies a technology roadmap for a strong smart grid, considered to be the first programmatic standard for guiding technology development in smart grid. By the end of 2015, smart grid benefits began to appear, meaning that national smart grid technology had reached worldwide level.

9.3 Current Situation of Foreign Countries' Smart Grid

9.3.1 United States

With focus on the construction of a smart power grid, the American government has put priority on the research and development of important technologies and embarked on formulating a development plan. The *2010–2014 Smart Grid R&D Multi-year Project Plan* is designed to set a holistic research and development project concerning smart grid to facilitate the development and application of technologies in the power grid by winning support from the community. According to the United States Standard and Institute of Technology, the work will be phased over three stages and a standardized framework of smart grid (i.e., 75 standard specifications, standards, and guidelines) is now released.

The US Silver Spring Networks Corporation provides power companies with smart grid-oriented setup and operating solutions for advanced ammeter framework (AMI). American Accenture is responsible for managing Boulder, CO's smart grid pilot project "smart grid city" and smart city projects in Amsterdam (Holland) and Yokohama (Japan).

In June 2013, the first large-scale smart power grid was set in motion in Florida, and Florida Electric Lighting Company was responsible for it. The smart grid system uses 4.5 million smart electricity meters and more than 10,000 instruments and equipment, and is characterized by networking between devices, thereby enhancing the flexibility and resilience of the power grid.

Millions of smart meters were put in use by November 2013, and the National Institute of Standards and Technology has a revised guideline for the installation and implementation of power grid technology so as to advance the construction of smart grid.

9.3.2 Europe

European countries carry out their research on smart grid and pilot projects by combining their technological advantages and characteristics of power development. Britain, France, and other countries focus on the development of pan-European grid interconnection; Italy puts priority on smart meters and interactive distribution network; Denmark focuses on the development of wind power and its control technology.

Britain has developed a "smart grid roadmap for 2050" and it supports research on and demonstration of smart grid technology. Construction will be completed in strict accordance with the specified road map.

In 2012, 27 EU member states and their associated countries – like Croatia, Switzerland, and Norway – had put 1.8 billion euros into research and innovation activities; 281 innovation projects related to smart grid research and development in total. Britain, Germany, France, and Italy are the four major investing countries among the EU members in smart grid technology application and development pilot projects, while Denmark is the most active in research, development, and innovation. The EU Seventh R&D Framework Programme (FP7) and 95% of EU-level innovation receive multi-country financial funding, along with close to 95% of cooperative research and development projects. FP7 plans to financially support programs mainly in the fields of two-way connection technology between consumers and power grid, efficiency increase technology, and ICT transmission grid application technology.

In January 2013, the German Federal Ministry of Economics and Technology, the Federal Ministry of Environment, and the Federal Department of Research jointly initiated a future attainable electricity network and specified that these financially supported R&D programs should be clearly limited to the power grid, with priority on the smart distribution grid, transmission network, and connection of offshore wind power, and solutions for associated interfaces, whilst also taking innovative research on energy-related systems analysis, standardization, and environment into consideration.

In February 2013, the European Committee for Standardization, the European Committee for Electro-technical Standardization, and the European Telecommunications Standards Institute developed standards for smart grid, smart meter, and electric vehicle charging.

A global cooperative agreement was made and entered into by and between French Alstom and America Intel in June 2013. Both parties are planning to work together in the fields of smart grid and smart cities, the development of related technologies and solutions, with the focus on embedded intelligent IT system security and a new grid frame launched for the future. The French Power Company planned to advance the construction of smart grid by virtue of smart grid standards. Denmark started a new smart grid strategy in 2013 to accelerate the implementation of consumer self-management of energy consumption. This strategy will introduce new meters counting by the hour, and encourage lower-price electricity consumers by means of a multi-stage tariff and data center. Currently, Denmark takes the lead in aspects of research, development, and demonstration of smart grid among the EU countries.

There is an ongoing study on optimizing smart grid at the Scottish Cumbernauld research center to raise power generating efficiency. As part of Scotland's smart big grid strategy, the center uses a micro-grid to test new technology.

9.3.3 The Americas

The Standards Council of Canada calls for a guiding committee in the smart grid standard roadmap to drive smart grid standardization and set policy and goals. The roadmap has been developed under the supervision of Natural Resources Canada and the Canadian National Committee under the International Electrotechnical Commission.

Eletropaulo, a Brazilian power company, announced in August 2013 that a wireless metropolitan area network would be applied in its smart grid project. This is Brazil's largest smart grid project. By 2015, the smart grid was due to have access to every corner of the São Paulo area and meet 60,000 households' demand for electricity.

9.3.4 Multinational Cooperation

The US S&C Electric, Samsung SDI, Britain National Grid, and German Younicos decided to jointly develop Europe's largest smart grid energy storage project in July 2013, located at a UK power substation in Leighton Buzzard, UK.

Nippon Electric Company (NEC) entered into an agreement with Italian Acea in February 2013, and both parties agreed to develop a lithium-ion energy storage system for installation at primary and secondary substations. NEC is expected to deliver two sets of energy storage systems and provide real-time monitoring of the status of charging/discharging as well as temperature. A new company was started in August 2013 with funds from Toshiba and Tokyo Electric Power Company, to engage in the study of smart meters, batteries, system applications, serving, and maintenance to expand the overseas business. After Toshiba purchased the world's largest smart meter manufacturer – Switzerland's Landis+Gyr – it conducted validation tests for summer dynamic response by use of a smart meter in New Mexico. The most widely used system is a grid control monitoring system, composed of an integrated management system of the Toshiba Group, a meter data management system, and a customer information management system of Toshiba Solutions Corporation.

The Smart Grid Association of Korea is launching a national program to encourage and support smart grid patent development in accordance with international standards. The association gives support to corporations, universities, or research institutes that apply for international patents. The US's ZBB Energy Corporation and South Korea's Honam worked together in some fields to optimize the manufacturing process of a 50 to 500 kWh V3 zinc bromine battery in January 2013. A ZBB storage system prototype will be delivered to R&D laboratories in South Korea, and the V3 battery will be applied to the Korean smart grid demonstration project.

9.3.5 EPRI USA Smart Grid Demonstration Initiative: 5 Year Update on Multinational Cooperation

The *5 Year Update* describes the development of foreign smart power grids from another perspective. As the report describes, Australia, France, Ireland, Japan, and America had once input tens of thousands of dollars into the power grid to realize investment and information sharing. Broad research results and goals are achieved in smart grid technology and demonstration applications. The pilot project finally integrated design, planning, and evaluation after 7 years of research efforts. This research focuses on the way distributed energy resources get access to the public grid and market operations after distributed energy is launched into the market.

1) *Illinois smart grid demonstration project.* Two distributed systems are demonstrated in the services of Illinois Ameren Corporation to reduce the impact of stepping down the voltage. The results of the test will become part of the technology deployment of Ameren's smart grid.
2) *American Electric Power smart grid demonstration project – voltage reactive power optimization software.* After the voltage reactive optimization system or community energy storage is in place, power grids will improve the control strategy (or modify parameters) to form new grid control technology.
3) *AEP smart grid demonstration project.* The AEP smart grid demonstration project is to evaluate distributed energy resources and technologies. The project will simulate a physical power plant, distributed power generation, power energy, and demand response system. The study is designed to evaluate the potential influences of smart grid technology on integrating resources. For example, when electric vehicles (EV), community energy storage (CES), and photovoltaic (PV) power generating systems are running in a distribution network, the project studies whether the EV dynamic

management technology and running systems will affect other problems of CES. For instance, it studies whether some technologies would improve the operation of another device or tap the potential, as well as whether a combined technology would affect other optimization techniques.

4) *First Energy project.* This project studies an integrated visual control platform. The project has a visual integrated control platform which provides information and an integrated view relative to the distribution system. The system provides real-time operation of a regional distribution network. The visual system records real-time information and saves it onto file for trouble clearing and planning. The four types of data are displayed directly through an integrated display unit, and they are respectively: load control equipment, distribution line sensor, substation equipment, ice storage device/peak load transfer.

5) *Hydro-Québec project.* This studies anti-islanding protection and control of distributed power generating systems.

9.4   Energy Network

Owing to the advancement of power electronics technology, information network technology, and automatic control technology, smart grid is characterized by high information, automation, and interaction, and has become more advanced and efficient, cleaner and greener, more ubiquitous, flexible, transparent, open, and user-friendly. Therefore, it is a trend to provide users with personalized electricity services to meet their diverse energy needs. A global energy internet proposes a sustainable solution to energy supply with a high view and higher strategic height, emphasizing clean energy replacement and global energy values. This global energy internet provides opportunities for the development of smart grid.

9.5   Opportunities and Challenges

Smart grid should seize the chance of a global energy internet, emphasizing both strong and smart, and realize innovation/make breakthroughs in energy, power grid, energy storage, and information communication.

1) *As renewable energy sources develop, the power grid should expand its acceptance capability.* Faced with a deteriorating global environment, gradual consumption of conventional petrification energy, and increasing costs of application, it is an inevitable choice to develop renewable energy sources – such as wind power and photovoltaic energy – to achieve sustainable development. Measures should be taken during power generation, transmission, and distribution to relieve bottlenecks, accepting the capacity of renewable energy sources to drive the development of new and clean energy.

2) *An increasingly complex power grid requires better control.* With the increasing size of power grids, the rapid development of new energy and distributed power supplies, and the advancement of integrated regulation/control, by prefecture and county, the existing power grid dispatching technology cannot fully meet the future needs of

operating power grids. This is because of the lack of the necessary ability to predict, regulate, and control wind, solar energy, and other intermittent power supplies; real-time or next-day dispatching technology is still in its infancy; the backup regulating system cannot fully satisfy the needs of regulation given the failure of the master regulator for long hours; the equipment and management levels are uneven, and badly in need of standardization.

3) *Growing demand for interaction between power grid and consumers.* In order to meet the needs of a distributed power grid, ensure reliable power supply, promote clean energy development, guide efficient consumption, and raise standards of service, there is an urgent need to establish a user-friendly service system, ensuring reliable power supply, promoting green development, and driving the development of correlated industries.

4) *The traditional distribution network needs to upgrade to a modern smart grid.* Greater demands are being placed on distribution network planning and design, access management, operation and maintenance, security coordination and control, so it is necessary to strengthen distribution network construction and reconstruction, and build a well-structured, advanced, flexible, reliable, economic, efficient, highly intelligent, open, and interactive modern smart distribution network.

Reliability requirements of the distribution network: economic development; transformation of mode of economic development; improvement of industry mix; high-tech industry; high value-added industries; high-precision manufacturing industry.

Multi-source feature of distribution network: various kinds of distributed power supply, electric vehicle, and energy storage, and other diversified power supplies with large-scale access to the distribution network; passive network transformed into active network; unidirectional power flow turns into multi-directional power flow.

5) *Power system equipment is facing great opportunities.* The world has seen the trend toward a global energy internet and "internet+" program. So, key technologies applied to green energy, smart networks, energy-saving applications, power generation, transmission, transformation, distribution, utilization, and scheduling, the development of smart equipment, and upgrading smart manufacturing are facing great opportunities.

References

1 Amin, M. 2001. Toward self-healing energy infrastructure systems. *JEE Computer Application in Power*, 14(1): 20–28.
2 EPRI. 2003. Power delivery system and electricity market of the future. Report 1009102, Palo Alto, CA.
3 US Department of Energy. 2008. The Modern Grid Imitative National Energy Technology Laboratory.
4 Rockefeller, G.D., Wagner, C.L., Liners, J.R., *et al.* 1988. Adaptive transmission relaying concepts and computational issues. *IEEE Transactions on Power Delivery*, 3(4): 1446–1456.
5 Jampala, A.K., Venkata, S.S., and Damborg, M.J. 1989. Adaptive transmission protection concepts and computational issues. *IEEE Transactions on Power Delivery*, 4(1): 177–185.
6 Sidhu, T.S., Baltazar, D.S., Palomino, R.M., *et al.* 2004. A new approach for calculating zone-2 setting of distance relays and its use in adaptive protection system. *IEEE Transactions on Power Delivery*, 19(1): 70–77.
7 Qin, B.L., Guzman, A., and Schweitzer, E.O. 2000. A new method for protection zone selection in microprocessor-based bus relays. *IEEE Transactions on Power Delivery*, 15(3): 876–887.
8 Ge, Y. 1997. Adaptive relay protection and outlook. *Automation of Electric Power System*, 21(9): 42–47.
9 Suonan, J., Xu, Q., and Song, G. 2005. Adaptive ground distance relay. *Automation of Electric Power System*, 29(17): 54–69.
10 Liu, Z., Zhang, Y., and Shen, Y. 2005. Research on adaptive ground distance relay. *Automation of Electric Power System*, 29(10): 21–26.
11 Ingelsson, B., Lindström, P.O., Karlsson, D., Runvik, G., and Sjödin, J.O. 1997. Wide-area protection against voltage collapse. *IEEE Computer Application in Power*, 10: 30–36.
12 Ren, X., Huang, J., and Huang, W. 2007. New approaches to schedule maintenance for AC/DC parallel transmission-line. *Proceedings of the CSEE*, 27(4): 65–71.
13 Wei, S., Xu, F., and Min, Y. 2006. Model for transmission-line maintenance plan. *Automation of Electric Power*, 30(17): 41–49.
14 Liu, J., Li, X., Liu, D., and Pan, J. 2010. Analysis of basic data platform for condition-based maintenance system of electric transmission and transformation device. *East China Electric Power*, 38(2): 216–219.

Self-healing Control Technology for Distribution Networks, First Edition. Xinxin Gu and Ning Jiang.
© 2017 China Electric Power Press. Published 2017 by John Wiley & Sons Singapore Pte. Ltd.

15 Bretthauer, G., Gamaleja, T., Handschin, E., Neumann, U., and Koffmann, W. 1998. Integrated maintenance scheduling system for electrical energy systems. *IEEE Transactions on Power Delivery*, 13(2): 665–660.

16 Wang, C. and Sun, G. 1998. Review on smart model research of fault diagnosis. *Hydraulic and Electric Machine*, 3.

17 Gu, X., Jiang, N., Chen, X., *et al.* 2008. About highly reliable urban power grid. *Power Supply & Utilization*, 25(4).

18 Gu, X., Jiang, N., Ji, K., *et al.* 2009. Implementation and outlook on self-healing control technology for smart distribution network. *Electric Power Construction*, 30(7): 4–6.

19 Chen, X., Gu, X., Yu, K., *et al.* 2009. Architecture of urban power grid self-healing control system. *Automation of Electric Power System*, 33(24): 38–41.

20 Guo, Z. 2005. Scheme of power grid self-healing control. *Automation of Electric Power System*, 29(10): 85–91.

21 Ren, J. and Guo, Z. 2007. Research on condition-based estimation model of power grid self-healing control. *Power System Technology*, 31(3): 59–63.

22 Gu, X., Chen, X., *et al.* 2009. Self-healing method for urban distribution network: China. Patent no. 10243972.8.

23 Su, C.L., Lu, C.N., and Lin, M.C. 2000. Wide area network performance study of a distribution management system. *International Journal of Electrical Power & Energy Systems*, 22(1): 9–14.

24 Tu, Y. 2014. Research and application of intelligent distribution network communication technology. Report, North China Electric Power University.

25 Huang, S. 2010. Business requirement analysis and technical scheme of intelligent distribution network communication. *Power System Communication*, 31(212): 10–12, 17.

26 Li, J., Yan, L., Qi, H., and Meng, F. 2014. Research and practice of private network of power wireless communication based on LTE230 system.

27 Zhang, Y., Yu, Q., and Zhao, L. 2013. Application of communication technology in intelligent distribution network.

Postscript

In May 2009, the SGCC released a remarkable report entitled *Building Strong Smart Grid*, unveiling this development project for the first time. Part of the new technologies for smart power grid had been explored and carried out in advance.

In 2009, the SGCC elaborated three stages in the development of the strong smart grid, based on the principles of integrated planning, distributed implementation, pilot projects, and all-round advancement.

1) Phase I between 2009 and 2011 – the pilot stage. At this stage, the pilot work for the key technologies of smart grid would be implemented all round by 2011.
2) Phase II between 2012 and 2015 – the overall construction stage. At this stage, the goal is to build a fundamentally strong smart grid, with key technologies and equipment reaching worldwide advanced level by 2015.
3) Phase III between 2016 and 2020 – the improvement and completion stage. At this stage, the goal is to build a comprehensive strong smart grid, with key technologies and equipment reaching worldwide advanced level by 2020.

Over the past two years, remarkable progress has been achieved in smart grid. The first trial projects have been fully completed, and secondary pilot projects are in the process of acceptance.

In the field of smart grid, a large number of achievements with independent intellectual property rights have been achieved in the process of research and construction of distribution devices, distribution lines, terminals, distributed power (micro-grid) connections, automation, monitoring distribution transformers, reactive compensation, automation circuit breakers, smart terminals, communication systems, and master station systems. Distribution automation systems have been built at the center of 19 urban areas. Power grid technologies have been rapidly upgraded and developed, after being put into practice.

The recent development of a smart substation has filled a gap in respect of research key devices and building standard systems. Currently, the SGCC has set 15 standards regarding smart substations, the world's first standards for smart substation series, and applied 126 patents with worldwide advanced technologies. Eight smart substations have been built in China, including Shaaxi, 750 kV Yanan, Jiangsu 220 kV substation, and Xian substation; these have played a vital role in upgrading the world's substations.

As the writer of this volume, I feel proud to be participating in the building of smart grid. The book is dedicated to the memorable era of smart grid, and to those colleagues – at home and abroad – who have worked hard for it.

Self-healing Control Technology for Distribution Networks, First Edition. Xinxin Gu and Ning Jiang.
© 2017 China Electric Power Press. Published 2017 by John Wiley & Sons Singapore Pte. Ltd.

Index

Self-healing Control Technology for Distribution Networks, First Edition. Xinxin Gu and Ning Jiang.
© 2017 China Electric Power Press. Published 2017 by John Wiley & Sons Singapore Pte. Ltd.